▶ERP

OPTIMIZATION

Using Your
Existing System to
Support Profitable
E-Business Initiatives

▶ERP
OPTIMIZATION

Using Your
Existing System to
Support Profitable
E-Business Initiatives

Cindy M. Jutras

CRC Press
Taylor & Francis Group
Boca Raton London New York

CRC Press is an imprint of the
Taylor & Francis Group, an **informa** business

CRC Press
Taylor & Francis Group
6000 Broken Sound Parkway NW, Suite 300
Boca Raton, FL 33487-2742

First issued in paperback 2019

© 2003 by Taylor & Francis Group, LLC
CRC Press is an imprint of Taylor & Francis Group, an Informa business

No claim to original U.S. Government works

ISBN-13: 978-1-57444-332-5 (hbk)
ISBN-13: 978-0-367-39561-2 (pbk)
Library of Congress Card Number 2002031871

Library of Congress Cataloging-in-Publication Data

Jutras, Cindy M.
  ERP optimization : using your existing system to support profitable e-business initiatives / by Cindy M. Jutras.
      p.   cm.
  Includes bibliographical references and index.
  ISBN 1-57444-332-1 (alk. paper)
  1. Electronic commerce. 2. Production management. 3. Management information systems. 4. Business planning. I. Title.

HF5548.32 .J88 2002
658′.05—dc21   2002031871

**Visit the Taylor & Francis Web site at
http://www.taylorandfrancis.com**

**and the CRC Press Web site at
http://www.crcpress.com**

DEDICATION

To my husband Glenn, who is truly my other half.
Together we are much more than we could ever be alone.

PREFACE

While businesses cannot operate effectively today without it, Enterprise Resource Planning (ERP) alone cannot meet the needs of businesses undergoing an E-transformation. Yet what company is not required to make such a transition? Today no company can operate in isolation or even at arm's length from customers, suppliers and partners. ERP forms a foundation for successfully meeting your E-business needs. But future success will be derived not only from a firm, supportive foundation, but also from a solid E-business superstructure to manage and guide it through the complexity of a value chain that has grown longer and more complex.

This book provides a technology-centric approach to guiding your company through E-transformation — an approach that leverages current investments and minimizes disruption to your business. But it does not require you to be a technologist, or even technology literate, to understand what you must do and the tools at your disposal to make it happen. Written in a business vernacular, it effectively bridges the gap between technology and business strategy.

Starting from the beginning of the transition, the book takes managers through the process of assessing their current position, and introduces them to the technology that can provide them with a unified view of their organizations, effectively leveraging all available information to provide true business intelligence. It details the steps they will need to take to assume a leadership position within an integrated business community, demonstrating how to support secure information exchange with customers, suppliers and partners. It shows them how to define interactions within and between companies, how to automate business process flows and most efficiently drive events throughout the value chain.

No, ERP alone is not enough to secure and maintain a leadership position in today's economy, yet it provides a backbone and an infrastructure of enterprise applications that are necessary, essential prerequisites to conducting business. Preserving and leveraging this value is the secret to success now and in the future. Whether you are a business executive today, or hope to be one in the future, this book offers a comprehensive guide for effectively building on the foundation of ERP in order to manage and guide a business successfully into the E-business world.

THE AUTHOR

Cindy Jutras is a seasoned software and business professional with over 28 years of experience in applying software solutions to business problems. Experienced in a wide range of functions related to the software industry, including sales, marketing, product development, customer support and product management, she is also an industry observer and trendsetter in business and business applications. Having worked with manufacturing companies for the full extent of that time, she is both a visionary and a pragmatist.

She is currently Director of Solutions Management for SSA Global Technologies, since their recent acquisition of interBiz, previously a division of Computer Associates. At Computer Associates she was divisional Vice President of Product Strategy and was instrumental in defining and guiding the product direction of ERP systems as well as advanced technology products. Prior to this, she held positions at a variety of manufacturing, consulting and software companies, where she played a diverse set of roles in operations management, research, design and development, as well as both product and project management.

A native of Massachusetts, Cindy graduated summa cum laude from Merrimack College with a degree in mathematics, with concentration in physics and computer science. She went on to receive a Masters of Applied Science from Boston University.

Ms. Jutras's work has been published and she is frequently quoted in industry publications on a variety of topics including manufacturing, ERP, E-commerce and E-business management, and customer relationship management. She is the original author of the concept of "Virtually Vertical Manufacturing" and speaks at industry events on this and other topics.

CONTENTS

1

CAN ERP MEET YOUR E-BUSINESS NEEDS?

Can ERP meet your E-business needs? This is a deceptively simple question with no singular, simple answer. Although businesses today cannot operate effectively in today's economy without it, ERP alone cannot meet the needs of businesses that are undergoing, or have undergone, an E-transformation. Which companies are those? Quite simply, the only companies that need *not* transform themselves are those that do not intend to be in business a few short years from now.

Not since the Industrial Revolution has the world experienced such intense and fundamental change. We are truly in the midst of a Technological Revolution, which has had more far-reaching effects than we could ever have imagined. Technology has caused the world to shrink, and global competition to become fierce. Businesses that once operated relatively autonomously at an arm's length from their customers and suppliers are now being forced to work more cooperatively and collaboratively within integrated business communities. Businesses, which only a few years ago operated locally, have expanded regionally, then nationally, and finally internationally, to the point where even modest businesses can be profoundly affected by changes in the global economy.

With this turn of events come global opportunities. And what opportunities there are! Yet, as always, with these opportunities come both challenges and risks. During the past two decades traditional business applications have evolved into integrated ERP systems. Many implementations have evolved as well, integrating both functions within a company as well as the business applications that support those functions. So inventory, order entry and shipping systems are fully integrated with financial accounts payable, accounts receivable and general ledger systems. Yet most ERP implementations stop there. Except for the occasional foray into EDI (Electronic Data Interchange), they focus primarily inside the business, and do little to connect externally to those integrated business communities. Yet that is what E-business is all about — breaking down the barriers of communication and interoperability between businesses, potentially on a global basis.

It is the accelerating pace of business and the corresponding quickening of the pace of change itself that forces us to look at ways to maximize the return on existing investments, with the least amount of disruption to business. So if ERP is not enough, what next?

TIME — THE MOST PRECIOUS COMMODITY

There is no longer a question of "if" a business will be transformed into an E-business, but more a question of how and when. The economics of E-business are compelling. E-business can provide new market strategies and fundamentally change the way you attract new customers and retain existing customers. It also provides some of the best opportunities to streamline business processes and reduce costs. Yet there will be companies that resist this transition, just as there have always been companies that lag behind and resist change. However, today the accelerating pace of business, combined with the accelerating pace of change itself, can easily turn such resistance into a death sentence.

Like it or not, technology is changing the world in which we live. The old maxim, "nothing is constant but change," has never had such relevance as it does today. Impelled by the rapidly changing technology that drives the world, this pace of change is accelerating to the point where what was considered "high tech" less than two decades ago now seems comically primitive. Today scientists predict that by 2020, we will attend videoconferences by slipping on a special pair of eyeglasses, wristwatches will access the Internet, and our clothes will be able to monitor heartbeats, alert ambulances and provide medical histories. What may have appeared far-fetched a few short years ago seems perfectly plausible today.

To place this rapidly accelerating pace of change into perspective, consider the history of modern communication, beginning with 1775 B.C. when the Greeks developed the first phonetic alphabet. The result of scaling the time, from 1775 B.C. to the recent turn of the century, into the equivalent of a 24-hour period, is shown in Table 1.1. The printing press has been hailed as one of the most important developments in the history of communication. Yet using our timescale, this did not occur until 8:30 P.M., and we see that the remaining truly significant technology advancements began just before the final hour of our day. The rise of the World Wide Web, which is changing the way we do business today so dramatically, occurred

Table 1.1 The History of Communication

1775 B.C.	Greeks develop first phonetic alphabet	midnight
1400 B.C.	Oldest record of writing (on bones)	2:22 A.M.
1140 A.D.	Cloth from mummies used to make paper	6:30 P.M.
1453 A.D.	Gutenberg prints 42-line Bible	8:30 P.M.
1838	Morse exhibits electric telegraph	10:59 P.M.
1876	Bell invents telephone	11:12 P.M.
1901	Marconi sends radio signal across the Atlantic	11:22 P.M.
1949	Network TV introduced in United States	11:40 P.M.
1951	Computers sold commercially	11:41 P.M.
1966	Xerox sells the Telecopier — a fax machine	11:47 P.M.
1980	CNN 24-hour news channel established	11:52 P.M.
1981	IBM PC introduced	11:53 P.M.
1985	Cell phones in cars	11:54 P.M.
1994	Introduction of World Wide Web	11:58 P.M.

only in the last 2 minutes. Think how far we have come in the final hour, and then consider the possibilities that lie before us as we progress further into the new millennium.

Many have compared the tragic events of September 11, 2001 to the tragedy of Pearl Harbor 60 years earlier. The surprise attack on Pearl Harbor came at 7:50 A.M., Hawaii time. The earliest report was 50 minutes later when a local radio station released the first news of the attack, and at 9:00 A.M. broadcasts spread across Honolulu. It wasn't until 3 hours after the attack that the mainland learned of the news, and it was 9 days later that the *New York Times* printed the first pictorial images. In contrast, the first hijacked plane hit the World Trade Center at 8:46 A.M. Eastern time. One minute later CNN began live national broadcasts. Four minutes later Associated Press sent out its first bulletin. Within minutes of the attack millions of Americans were watching the events unfold on televisions in homes and at work.

After December 7, 1941 it took days and even weeks for relatives of survivors to find out if their loved ones were safe. Although long distance calls were disrupted immediately after the September 11th tragedy, dial-up and networked Internet connections brought instant relief to many. My own friends, family and associates know how often I fly out of Boston's Logan airport, and that I typically spend 2 to 3 days in New York each week. E-mails started pouring in to me, all with the same questions: "Where are you?" and "Are you OK?" Fortunately, nobody had to worry about me for very long, as I sat in my office and replied. But, sadly enough, instant connections played an even more important role as cell phones connected victims to their families for last good-byes.

However, let's bring this down to more commonplace day-to-day events for the moment. When you communicate a question today, either personally or professionally, how quickly do you anticipate a response? Look at how the process of communication has accelerated. In the 19th century a Bostonian could write a letter to Hawaii and expect a reply in about a year. When sailing ships were replaced by the U.S. mail service, it still took weeks for a letter to be delivered across the Pacific. Even as recently as 30 years ago, first-class U.S. mail had a 1-week lead time, when operating coast to coast, and that was considered good. When Federal Express first guaranteed overnight delivery across the country it was a premium service and used only for the most valuable or perishable shipments — certainly not for anything as mundane as a letter. Then there was the fax machine, and now the Internet. Could a sailing captain in 1830 delivering a letter to Hawaii have imagined overnight service? E-mail and Internet chat rooms? Today "overnight" is the slow way of communicating. I know people who turn their e-mail "Out of Office Assistant" on when they will be off-line for several hours — because some of us have come to expect almost instantaneous response. We carry cell phones, laptops, pagers and PDAs (personal digital assistants). New area codes are popping up everywhere to handle increased demand for telephone numbers. Whether you view this as good news or bad, we are literally never out of touch.

Why? Because things can't wait. We are impatient. We demand instantaneous response. Society demands instantaneous response. Business demands instantaneous response. Entire industries demand instantaneous response. It's hard to imagine how the world can get any faster. But it will. And those companies not able to respond will be left in the dust.

Time is quickly becoming one of the most precious commodities in today's intensely competitive markets. Technology advances and planned obsolescence make lead time to market a critical success factor. Dominant players in selected markets today continue to push risk downstream, demanding faster delivery and more responsive service. These industry giants don't pad their inventory in the event of exceptional demand. Instead, they expect their suppliers to respond instantly.

These demands call for successful companies to be exceptionally agile, able to respond quickly and efficiently, both to changing markets and economic conditions in a global environment, and also to exceptions and disruptions in day-to-day operations. The net result is that businesses cannot afford simply to respond to the next technological innovation, but they must learn to respond to change as a constant state. As difficult as it may appear, it is not enough to react to the E-business challenges the World Wide Web is presenting today. Companies must position themselves to be able to respond to and take advantage of the next technological advance, whatever that opportunity may be.

How have companies to date responded to the demands created by this accelerating pace, and still maintained the profit margins necessary for them to function and grow? Interestingly enough, many of the challenges faced today by businesses are both caused by and enabled by technology. It is technology that enables the automation of once manual processes. It is also through technology advances that products are made obsolete at an unprecedented pace. Technology causes the world to shrink, and global competition to become fierce. As a result of this competition, many companies have been forced to downsize, leaving leaner, flatter organizations. Fortunately, the reduction in human resources can often be offset with additional technological resources. Both plant automation and information systems technology allow companies to do more with less. While this has interesting implications for the social fabric of our culture, without these changes, many companies today would not survive.

Another response to the challenges facing enterprises today is to reengineer business processes. Most business reengineering efforts target functions that add no value to actually getting product or services out the door. Some of these functions can be eliminated entirely, while some are necessary support activities that add indirect value. Invoicing may be eliminated entirely with customer self-billing procedures. Small purchases may be handled by way of credit cards instead of purchase orders. Common functions, such as sales order entry, while adding no direct value to the product being delivered, are necessary steps in the business process. Some companies would benefit from having someone else's employees (e.g., their customers'), or consumers themselves, perform this function. These types of initiatives are changing the way companies do business and, in turn, changing the relationships forged between trading partners.

While ERP today supports each of these business processes, does it provide the necessary flexibility when the control and the very nature of the process shift?

"VIRTUAL" INTEGRATION

The strongest companies in the past were those that were vertically integrated. By being self-sufficient through each step of the entire process of transforming pure raw materials into a finished consumable good, and taking the added responsibility of delivering that product to the end customer, the strongest chain has been forged. By having a single company responsible for all of the steps in the entire process, that company has the greatest possible control, the least contention for materials and the best chance of synchronizing all the necessary components of the process.

However, there are inherent weaknesses in this approach. While it reduces tension between the links, this is basically the most rigid of supply chains. Because no company can achieve excellence in everything, this approach is also the most likely to produce a weak link and the most susceptible to failure as a result. And if that weak link fails, there are fewer alternatives that can be immediately implemented in order for the process to maintain the speed and integrity of the value chain.

This has caused companies to evaluate product offerings and business processes carefully to determine the potential for outsourcing. Whether it is through the purchase of subassemblies or finished product, the contracting of manufacturing or distribution services or the outsourcing of customer services or information technology, the goal is to create a more efficient, responsive and flexible environment.

While companies have tended to become less vertically integrated in attempts to focus on their core competencies, this also has necessitated, and will continue to necessitate, new ways of companies doing business with each other. The value chain has lengthened and become more complicated; yet expectations of response time and delivery performance have risen dramatically. These new business relationships, depicted in Table 1.2, are established in an attempt to recreate the ability to control the entire process while spreading and reducing the risk factor. As a result, the most successful companies are those that fully participate in and demonstrate leadership within an integrated business community.

Table 1.2   Traditional vs. Virtually Vertical Manufacturing

Traditional Manufacturing	Virtually Vertical Manufacturing — Full E-Business Integration
Preferred suppliers with approved alternatives; fixed make/buy decisions	Interchangeable sources of supply
Buffer inventory reduces tension between organizations and inflates working capital	Lean, just-in-time inventory reduces working capital
	Supply chain-wide planning
Supplier sees immediate orders only	Shared business goals and forecast
	Collaborative planning and scheduling
Standard order to cash cycle; purchase order/sales order/invoice/cash	Point-of-use consumption
	Vendor-managed inventory
	Consignment inventory
	Self-billing procedures
	Credit card purchases
Traditional payment methods	Electronic funds transfer
Electronic Data Interchange (EDI)	Electronic Commerce
	Internet enablement

These new business models involve multiple companies working cooperatively and collaboratively together, in a seemingly seamless manner, as if they were a single, *virtually* vertical enterprise. A company that can successfully interoperate in this way can claim to have reached the goal of full E-business integration.

As a result of this push toward full E-business integration, businesses face challenges that force them to push the envelope of business information systems. ERP grew from its predecessors of MRP (material requirements planning) and MRP II, constantly expanding its solution footprint to address more and more of the needs of the enterprise. Yet ERP was not conceived to look beyond the "four walls" of the enterprise, regardless of how expansive those walls would become, simply because the concepts of MRP and ERP were born in a time when companies were run as independent enterprises with arm's length relationships with customers and suppliers.

Not so long ago, if you delivered a good product on time, at a reasonable price, and you paid your bills in a responsible time frame, how you managed your company was your own business. And ERP was designed to help you manage that business. With a successful ERP implementation you could process orders, and manage inventory and production schedules to produce the product within lead time. You could invoice customers, pay suppliers and balance your books. How automated your processes were was your business alone. Whether your information was managed using clerks and filing cabinets or online systems was transparent to your customers as long as you provided them the care and service they demanded. The fact that few companies fully exploited the powerful capabilities that led them to purchase their ERP system in the first place was a sad reality but an easily kept secret. Until now.

E-BUSINESS EQUALS FULL EXPOSURE

As more and more integrated business communities emerge, companies will find that what used to be a private issue is now a public concern, at least in the eyes of an integrated business community.

I was recently at a manufacturer in the apparel industry that was evaluating whether to replace its existing ERP solution or to consider alternatives that would leave it in place, but "technology enable" it. One of the directors expressed a concern over implementing a new ERP system because of a recent experience with one of his suppliers. In the past the manufacturer had been able to quiz this supplier on its ability to respond to a previously unanticipated demand. With its previous systems the supplier had been able to respond within a day. However, their new system required an order to be placed to do any kind of "what if" analysis, and therefore they could no longer respond to these speculative inquiries.

What should have been strictly an internal limitation of their new ERP system (or perhaps a training issue with the user) has now become a concern of their customer. The end result for the supplier will probably be the loss of this manufacturer's business, simply because it will find a supplier that is easier to do business with.

As you approach E-business, whether you do so eagerly or are dragged there kicking and screaming, you will find your business much more exposed. Experts in E-commerce and E-business routinely caution executives to reexamine business processes before blindly electronically enabling them, because doing so is the equivalent of throwing open all your doors and windows to public scrutiny. However, continuing to use this as an excuse for lagging behind where you know you need to be is just as dangerous.

The inward focus of ERP and its predecessors has had an insular effect on companies. It has been one of the barriers some companies have erected to protect themselves from the reality of our changing world. And they have survived; some have even prospered by simply focusing on producing a high-quality product, at a reasonable price, delivered on time. But most of these companies are already feeling the pressure of our changing economic climate. They are finding that standard products, which they can accurately forecast and stock on their shelves, no longer necessarily meet the needs of their customers. They have found that 13-week lead times are no longer acceptable, or even 10 weeks, or 3 weeks. Customers are becoming more demanding. While sales may remain strong, their ability to deliver begins to suffer. They are feeling the effects of our rapidly changing world, in spite of the barriers they have erected to protect themselves from this reality. What happens to these companies as their customer or even major suppliers become more demanding and start to require some forms of interoperability as a condition of doing business with them? How will they react to this change? Especially when competition can spring from any direction today.

Back a couple of years ago, speaking at a business conference in Sydney, Australia about the concept of virtually vertical manufacturing, I had a member of the audience challenge me with a question. He said, "I have too many problems internally to start collaborating and interoperating within a virtual enterprise. I'll never get to that point."

I asked him when his home was the cleanest. Of course, he admitted that it was the cleanest just before company was due to arrive. He also admitted to knowing some people who never cleaned their houses unless company was coming. And those were the very same people who were most likely to shut the doors to some of the rooms instead of getting them in order.

My point? Sometimes you need an excuse to get your house in order. So maybe your inventory accuracy in general is an embarrassment. So maybe you don't use collaborative planning or replenishment techniques with all your customers. But when your largest customer suddenly requires you to support same-day shipments, doesn't it make sense to clean up that room in the house first, leaving some other rooms closed for now?

BEYOND "QUALITY–PRICE–DELIVERY" TO INTEROPERABILITY

ERP forms a foundation for successfully meeting your E-business needs. For years companies have judged other companies on the three basic performance indicators of product quality, price and delivery. Therefore, every robust ERP system has means of measuring supplier performance within these categories. The same ERP systems are also able to measure a company's own performance in meeting the needs of its customers. However, there has always been another, less quantifiable concern that has subtly worked its way into business-to-business relationships. How easy is this company to do business with? In our electronic age, as the way businesses do business with each other changes, we will start to see another measurable performance indicator emerge — one that has largely been ignored by ERP implementations. The mantra of "quality–price–delivery" will soon be expanded to include a fourth metric of "interoperability."

Successful orchestration of the complexities of a modern business enterprise, particularly in the context of an extended, or virtually vertical enterprise, will be dependent on a company's ability to share information promptly, securely and effectively with customers, partners and suppliers. This efficient exchange of information will be required for organizations to become full participants in the collaborative planning and execution environments that drive the strongest value chains. This communication challenge, as it is largely, if not exclusively, electronic, is further complicated by a perverse diversity of standards. Yet in order for interoperability — i.e., how easy a company is to do business with — to be a viable condition of vendor selection, it must eventually emerge as a measurable standard. And speed and flexibility will be at least two of the components in the equation — how quickly you respond to requests for information and how flexible you can be in providing it in the format desired.

Through the process of E-transformation, companies must focus attention outwardly without losing any of the gains their inward focus brought them previously.

THE EFFECT OF E-BUSINESS ON YOUR BUSINESS

The need for speed and flexibility has been just one by-product of the Internet age. Today, the Internet allows us to reach more people in more places in a time frame that was previously inconceivable. Where it once took mature companies 30, 40, even 50 years to penetrate new, international markets, today a tiny start-up can establish a global presence on the Web virtually overnight. But with that global opportunity comes global risk. That presence, to the casual Web site visitor, is virtually indistinguishable from that of its well-established competitor, flattening the advantage of size and previous market presence. Therefore, the "dot-coms" of the day present a twofold risk. Even as a market leader, traditionally a "bricks and mortar" company, you may suddenly find new and unexpected competitors threatening your market share. But the risk extends even farther. The ability to market and sell to expanding markets can easily exceed a company's ability to fulfill the demand along with the expectations that are generated.

What does this mean to your overall business strategy? Certain functions within your business have the most potential for experiencing the impact of E-business. This is true whether you actively participate in E-business initiatives, or if E-business is simply being conducted around you.

THE SELL SIDE OF E-BUSINESS

The first of these functions is sales. The Internet can be viewed as an alternative sales channel and you will need to construct a strategy that reflects the possibility of channel expansion. For some companies the decision to sell product over the Internet or not will essentially be made for them. Today, when the decision is made outside your own executive suite, it is usually in the form of denial. Many large retail giants prevent you from competing against them directly with an electronic storefront. If these major retailers represent a large portion of your current or potential business, you will find the decision an easy one. You value their business enough to preclude competing against them in any manner.

However, many businesses today will be confronted with this decision. Will you sell direct over the Internet, and if so, what are your goals and how do you avoid channel conflict? In many ways, this basic business decision is no different from when only traditional choices are considered — for example, the decision to augment your direct sales force with a distributor network or other similar indirect channels. How do you do that and avoid cannibalizing your existing market? In many cases, distributors are recruited for some added value, allowing you to reach a segment of your market different from the one you have previously been able to satisfy. Can an electronic storefront offer you the same benefit?

Certainly one of the most compelling reasons for E-business initiatives is the potential for market expansion. Traditionally, there have been three ways to expand your market — through brand new product introductions, through new packaging and marketing of variations of your existing products and through geographic expansion — the opening of new territories. It is through this latter option that the Internet offers the most compelling advantages, particularly if there are sparsely populated or remote territories that were not economically feasible to cover with a direct sales force and expansion opportunities through indirect channels were limited or non-existent.

On the surface, the cost of opening these new "territories" might appear to be negligible, buried in the fixed cost of the information system investment of opening and maintaining your electronic storefront. However, you cannot ignore the added cost of marketing to these new territories to make your brand and your Web presence known. The World Wide Web is no "field of dreams" where if you build it they will come. Even in the world of E-business, instantaneous market penetration only exists on television commercials.

However, once you replace or supplement your direct or indirect sales force with an electronic storefront, you relinquish a certain level of control over your growth. With a direct sales force you limit growth by exercising control with your carefully planned recruitment efforts. The same can be said in terms of recruiting distributors. However, once you establish a Web-based sales presence, you are less able to gate that growth, and without new or revised methods of forecasting demand, you may well be caught in an out-of-control situation.

Your electronic storefront not only exposes you to increased and unplanned-for demand, which is a good problem to have, it could also potentially change the profile and ordering patterns of your customers. If you previously supplied large retail chains at a regional or corporate level, you may now be dealing with individual stores or the end consumer. Your order volume could go from one order per region per season to one order per store per week. The corresponding increase in order and transaction volume could very well be one or several orders of magnitude. Unless you plan accordingly, this could bring your current order entry system to its knees.

The selling and order administration processes are only the front office considerations for your entry into the E-commerce marketplace. There are also back-office considerations, most specifically the impact on your planning, warehouse and distribution efforts — in other words, fulfillment. Everyone has either a funny story or a horror story to tell from the 1999 holiday season. Gifts ordered in November that arrived in January. The two-line item sales order that arrived in two different shipments. One had your name spelled correctly; one didn't. What does that tell you about the integration of their order system with their fulfillment system? Holiday gifts that arrived in February. Products returned and never replaced. Credits issued only after multiple phone calls. Holiday gifts that arrived in July. Not being able to exchange the item you ordered online at the local retail store. Holiday gifts that never arrived.

As the profile of your typical order changes and your order volume increases, this will have a ripple effect throughout your fulfillment process. In the case where large retail customers previously ordered in aggregate and now order on a store-by-store basis, consider the impact on your forecasting and demand planning. You now have to deal with the demographic and location-specific variability previously smoothed by the aggregation and consolidation that were previously performed by the retailer. This can have a significant impact on your demand planning and your subsequent ability to respond to customer demand without a significant increase in inventory levels and the corresponding expense. Delivery to multiple locations can also have a very significant impact on your delivery mechanism, typically one of the first functions to be considered for outsourcing. Yet if you go from performing your own delivery to using other public or private carriers, what impact can this have on your business system requirements and, in turn, on your ability to manage this expense?

Now go to the next step and consider the impact as you add your own electronic storefront to the mix. You could easily go from shipping pallet loads to picking, packing and shipping individual items. Not only does your transaction volume significantly increase, but your warehousing strategy may very well need to be reengineered allowing you to maintain and transact inventory in any variety of units of measure — still more considerations for your ERP system.

THE BUY SIDE OF E-BUSINESS

Expanding your selling efforts is not the only E-commerce transaction you need to plan for in formulating your E-business strategy. The purchasing or procurement modules of ERP systems have long been modeled after how purchasing departments have operated for decades. Purchase requisitions go through a series of approvals. The automation of this requisitioning and approval process will vary between companies and between ERP systems, but ultimately the requisitions are funneled to the buyers of the purchasing department. Buyers are not only involved in vendor selection and negotiation, but also in the placement and administration of purchase orders, as well as monitoring vendor performance. While it is the first set of tasks — vendor selection and negotiation — that adds the most value to the enterprise, typically it is the function to which buyers are able to devote the least amount of time.

In an effort to automate and streamline the administrative efforts and to enforce spending policies more carefully, more and more companies have turned to Internet procurement. Internet procurement or E-procurement allows your employees to initiate purchase requisitions through an automated approval sequence that ultimately connects directly to approved suppliers with little or no manual intervention. Resultant potential for savings per purchase order has been nothing short of dramatic. ERP has made great strides in integrating the MRP and purchasing functions, resulting in automating purchase replenishments of direct materials. Therefore, most E-procurement initiatives begin with maintenance, repair and operational (MRO) or indirect purchases.

I have heard companies quote any number of estimates of the cost of processing an indirect purchase order, ranging from $50 to $350 to "I don't know." Some of these numbers are truly frightening considering the cost to process a purchase order for a $25 hammer can sometimes cost the same as processing a purchase order for a $300,000 software package. Yet once the technology infrastructure supporting Internet procurement is in place, these costs can drop to pennies per transaction.

The National Association of Purchasing Managers estimates that a third of indirect purchasing is noncompliant with company policy. But rarely do personnel deliberately ignore policy. They are simply finding expeditious ways to obtain items faster and more conveniently. Effectively implemented E-procurement efficiently controls access to items, pricing and suppliers, making fully compliant, online purchases easier and faster.

Several factors will determine the ability of your ERP system to handle this modified business process. The first and most obvious requirement of Internet procurement is whether your purchase requisition and purchase order system is Web enabled. Second, what level of training is required to use the system? How truly intuitive is interaction with the application? Is the process sufficiently automated through the use of workflow technology to eliminate manual intervention, which adds to the cost you are striving to reduce? And finally, consider the impact on your system when instead of 2, 5, 10 or even 50 trained buyers using it, you unleash potentially hundreds or thousands of employees as casual and occasional users. This has both security and performance implications.

And now let's bring this full circle. The net effect of this process is to place orders automatically with your suppliers. Ultimately, this means an order is captured in your vendor's order entry system, placing demand on its fulfillment processes. You must ask the same question of your vendors' systems as your own. First of all, how open and compatible are their back-office systems? This has a direct impact on their ability to interoperate with you. And also, has their ability to market and sell to expanding markets exceeded their ability to fulfill the demand and the expectations you have of their performance?

These and other considerations are discussed in more detail in Chapter 5, where we specifically define how your E-commerce strategy and the implementation of E-commerce applications can have a major impact on the ultimate goal of full E-business integration. That chapter will also discuss the different business models available and the interdependencies among these models, your business processes and your information technology.

SUMMARY

As you can see, as we enter into the world of true E-business integration, the opportunities are boundless for increased global market penetration, streamlined operations and reduced costs. But those opportunities do not come by accident, nor are they without cost or risk. The speed of business, as well as the speed of change itself, has made time the most precious of commodities today. Companies are forced to become more agile and more flexible as more and different demands are placed on them. No longer is the combination of price, quality and delivery sufficient. Today interoperability within integrated business communities and virtual enterprises is the new fourth dimension of success.

ERP systems have provided companies with a backbone for managing internal business processes and controlling transaction level activity at an arm's length from their suppliers and customers, but now businesses must take the next step, shifting their focus outward. Are you ready? E-transformation is not a single giant step, but a series of smaller steps that become a journey to full E-business integration. The following chapters are presented with the intent to help you focus on the steps necessary to exploit these opportunities, avoid obstacles and minimize your risk through the intelligent application of technology.

2

THE EMPOWERED EXECUTIVE — KNOWING @ THE SPEED OF E

Many C-level executives and senior managers today are content to remain distanced from the information technology that supports their businesses. But today you cannot separate business issues from technology, and becoming technology literate is just as necessary as being able to read an income statement and balance sheet. If you rely on subordinates to prepare and translate information for you, you run the risk of fragmented knowledge and an incomplete view of your business. Yet if you connect at the wrong level, you run the risk of opening the floodgates and being inundated with more information than you can possibly process and digest. This chapter describes the technology you need and how you must apply it to leverage the information assets available to produce a unified view of your business, create a management-by-exception environment, and bridge the gap between technology and business strategy.

A STORY

Back toward the later part of 1999 I was at an airport with a few minutes between connecting flights and made a quick stop in an airline club to check voice-mail and download my e-mail. The club was crowded but I found a free desk, hooked up and dialed in. At the desk next to mine was a gentleman speaking rather loudly into the phone. It was apparent to me (and probably to everyone in the room) he was speaking to his assistant or secretary, dictating messages and spitting out instructions. In fact, the volume of his voice made it rather difficult to listen to my own voice-mail messages, so as soon as I was finished downloading I packed up and headed to the gate. I had been fortunate enough to upgrade so I boarded early and settled into my seat. I thought I might even get luckier and have the seat beside me remain empty when, just before the doors closed, someone dashed on and dropped a briefcase onto the seat beside me.

Imagine my dismay when I looked up and found Mr. Loudmouth himself stashing his bags and dropping heavily into the seat, a phone still at his ear and a conversation still in progress that I was sure the entire plane could now hear. Feeling rather annoyed, I was relieved at our announced imminent departure, knowing at least the phone would have to be turned off.

Soon we were wheels up and my neighbor pulled out a folder at least an inch and a half thick and began reading and making notations on each piece of paper he read. As he furiously scribbled away I glanced over, curious to see what he was doing. Each page in his stack was an e-mail message, most likely one his secretary or assistant had printed for him. On each he would either place a large check, no doubt to mark that he had read it, or write a response, along with notes and instructions, again to a secretary or assistant who would presumably be the person who would actually reply to the message.

Seeing nothing particularly juicy, I pulled out my laptop and booted it up. As it started chirping through its start-up I noticed this gentleman surreptitiously glance my way. I'm not generally inclined to strike up airplane conversations so I ignored his interest, much as he had ignored mine. However, as several minutes passed his interest became less and less subtle and harder and harder to ignore. Finally I acknowledged his presence and he abruptly asked, "What are you doing?"

Still disinclined to striking up a conversation, I simply replied, "e-mail." He looked again, this time poking his nose right in front of my screen. He peered over his reading glasses for a few annoying seconds, during which I had a strong urge to snap my laptop shut. Finally he retreated back to his own stack of "e-mail." To be polite, I reciprocated and asked, "And what are you doing?"

He replied with a somewhat superior air, "Well, I'm doing e-mail, too, but I have an assistant who does the typing."

I couldn't resist. I smiled what I thought was my sweetest smile and asked, "And do you find that more efficient?"

I won't bore you with the details of the rest of our conversation. Although if you happened to have been on the same plane, I'm sure you heard it — at least his side of it. My neighbor, as it turned out, was the president of a medium-sized manufacturing company. He had been with the company for 28 years and president for the past 8. He assured me he had successfully avoided "newfangled gadgets," as he called them, until about 2 years ago. This was when his parent company had initiated some global standards. Among these standards were a new telephone system with voice-mail and also an e-mail system. Despite that he seemed to have taken well to that newfangled advancement of a cellular telephone, these changes were a little too close to automation for his taste, although he said he was "adjusting." But what had him extremely worried was a new initiative to replace his existing homegrown business applications, which were not Y2K compliant.

We talked a bit about why he was reluctant to give up these legacy systems. He didn't have any real, substantive reason beyond, "Those reports have been enough to run our business for the past twenty years, I don't see why we need new ones."

When I asked him what he liked best about these reports, he admitted he didn't really look at them. In fact, he never saw them at all, but his management staff used them and reported back to him. When I asked him how quickly they were able to obtain information about something happening, like the fulfillment of a customer's order, his reply was something akin to "my staff has what they need for our monthly meetings."

Then he pointed to my laptop and said, "Some of my people have been asking for these things, too. Pretty expensive toy, if you ask me." Of course, that struck a bit of a nerve with me since I tend to carry my whole life on my laptop. So I felt compelled to show him a few of the things I did with this "toy" — simple things like papers and presentations — no need to scare him! He shook his head and said, "My people don't write papers or give presentations."

"Well, what about letters, contacts and other correspondence?" I asked.

"We have secretaries to do that. My people's time is too valuable to turn them into typists. Besides, what's wrong with good old-fashioned pencil and paper?" he replied.

I could tell this was a company president that wasn't interested in turning on a dime. I shook my head and went back to my e-mail. And there we sat for the next 3 hours, side by side, each of us in our own worlds, about a million miles apart.

WHAT DO YOU NEED TO KNOW?

Do you know someone like this? Do you yourself perhaps feel some empathy for the fellow? I would venture to guess we all know someone who "doesn't use the computer." And I'm talking about business associates here. My 87-year-old mother doesn't use a computer, but I find that quite reasonable. A computer just doesn't seem to fit in her world. She doesn't do research. She doesn't own a credit card, so she couldn't purchase anything online. She writes beautiful letters and, as a former elementary school teacher, is one of the few people in the world today who thinks good penmanship is an important skill. I thought, very briefly, about getting her a computer and e-mail. But then I thought, who would she send e-mails to? Me? Widowed and living alone, she would much rather talk with me several times a week than send and receive electronic messages. And none of those lucky enough to get a weekly letter from Mom, most of them octogenarians themselves, has a computer. It almost made me wonder how the telephone ever really caught on. What good would a telephone be if none of your friends had them? But then a telephone had the very practical purpose of connecting you to emergency assistance even before teenagers discovered chatting for hours on it with their friends.

But yes, everyone knows someone like my fellow passenger — those who have their secretaries print their e-mail and respond to it in the same way, thereby quite effectively eliminating the real value it could bring them. Think about it. You send this executive an e-mail on Friday. His secretary prints his e-mails every morning of the workweek, so it is Monday before it gets printed. Maybe he reads them and "responds" on Tuesday, but he happens to be traveling and doesn't get back into the office until Thursday. He has also probably brought a lot of additional work for the secretary, so if we're lucky, she gets around to following up and typing in his responses on Friday. That's a week's lead time for something that could easily have been a simple request satisfied in a matter of minutes. I try to avoid e-mailing this kind of person because, I confess, I am too impatient for the response. But inevitably travel and time zones prevent you from connecting or you need to send a document to review.

Without further belaboring the point and beleaguering this poor individual, let's get practical here. While e-mail is just one example of technology, and a relatively simple one at that, I find executives' attitude toward this simple technology to be fairly indicative of their knowledge and their ability to tap into the potential of technology to help them run their businesses. Those who understand the value of timely communication also acknowledge the accelerating pace of business in our new and evolving economy. The need to become more flexible and agile in our mode of communication is just the tip of the iceberg when it comes to operating within today's business community.

In a research note published in April 2001 by a well-respected provider of research and analysis on information technology (IT) industries, I read, "Most business leaders are technology innovators, agile in both their vision and their execution." I doubt, however, that the research analyst would have used these terms to describe my fellow passenger. But then sometimes you have to distinguish between a titular head of a business and a true leader.

In fact, the senior-level management I have had the occasion to deal with directly have spanned the full spectrum of technology literacy — from my fellow air traveler to high-tech wizards who spend more time and money acquiring the latest technology than they do running their businesses. Obviously there are hazards to be negotiated at each end of the spectrum, but executives who do not have a basic understanding of the possibilities that are available through the effective implementation of technology put themselves, and their companies, at risk of not achieving their full potential. You don't have to understand how the technology works. In many cases you need not understand how to use it, although I would argue that some basic desktop skills can significantly enhance your effectiveness!

But what about beyond this? Is it enough to be buzzword literate? Where do you start? How far do you go?

TRANSACTION-BASED SYSTEMS A START, BUT NOT ENOUGH

The technology you need to run your business today can fall into multiple categories. The technology that supports basic voice and message communication within your enterprise falls into the category of basic infrastructure. This is also true of the technology you use to allow your employees to connect to both the Internet and to the information systems that are at the heart of your business. These are your telephone systems, your e-mail systems and the networks that support them and your business. If you don't think of your telephone as "technology," then your employees are probably not communicating in the most efficient and cost-effective way. However, these kinds of systems, in themselves, I will leave as topics for other authors. I will view them only in the context of if, how and when they need to be connected to that technology I include in the category of Information Management.

Any company of considerable size today has made some level of investment in information management. Whether you have a fully integrated ERP system implemented today, or simply run individual business applications that are interfaced loosely or not at all, these elements of IT are integral to running your business. These systems record the volumes of transactions that *are* your business. The buy and sell transactions. The movement of inventory. The flow of cash. They record the information about all your assets, including the physical assets of inventory, cash and equipment, as well as the human resources of time and intellectual capital.

Adequately leveraging your assets today includes the leverage of your information assets.

The amount of information collected in existing business systems today is colossal. Management is literally sitting on a gold mine of information. Yet when they need to make well-informed, strategic business decisions, few leaders have historically turned to their internal information systems for answers. Even fewer senior managers are actually directly connected to the applications that support their businesses. Why is that?

First of all, executives are typically not comfortable with technology. While many are more technology literate than my fellow passenger, the majority is still intimidated by it. But more importantly, there is simply too much detail. Most ERP systems today excel at producing mountains of information — so much so that it is impossible for a senior-level manager to plow through it, with any kind of efficiency, in search of answers. In addition, traditional business applications have excelled at collecting historical data from internal sources. But leaders today cannot drive their businesses on historical data alone any more than the driver of an automobile can navigate by only monitoring the gauges on the dashboard and looking in the rearview mirror. Instead, drivers must also look ahead, collecting data from many sources, to anticipate conditions in the future. And this highlights the third reason executives don't turn to these systems. They simply don't tell the whole story.

Much like passengers in an automobile, the welfare of shareholders today is dependent on factors both internal and external. The most careful driver today is still at risk on the road because of conditions on the road and because of what other drivers do or don't do. Whether it's a pothole in the road or another car that seems to come out of nowhere, the higher the speed the car is traveling, the more likelihood there is for a mishap, and in the event of an accident, the more devastating the crash will be. Driving any business today is far more akin to barreling down a super-highway than it is to meandering along a country road.

CEOs have always had to balance caution with risk. But typically today they are forced to drive their businesses at a much higher rate of speed than was required in the past. Throwing caution to the wind and growing too fast or being overaggressive in promising delivery can put stress on your business much as driving too fast can put undue stress on the engine of the car itself. As you accelerate you need to know that there are ample warning indicators that can signal danger. And as business leaders operate in this dangerously high gear, they must be aware of those potholes or other drivers as constant threats. These are the external factors, which are outside of their control, which their internal information systems will never identify. These may include demographic, political and economic factors, changing government regulations, and competitive information or natural or man-made disasters. Much of what affects strategic decision making happens outside the four walls of your enterprise.

IT'S NOT JUST ABOUT INFRASTRUCTURE

As a business leader it is natural for you to put the infrastructure in place that allows you to operate your business efficiently. That infrastructure includes the people you need, whether they contribute directly or indirectly to producing and shipping product. It also includes the equipment you need to make your product and generally to support your operations. And it includes that basic infrastructure technology for communication within and to the outside. And finally, you must put in place the necessary information management systems, although you may be more likely to think of these as a backbone of your business, as opposed to strictly infrastructure.

But there is another structure that you are less likely to see as a business necessity, and that is a management superstructure. While an infrastructure generally lies beneath and provides support to your business, a management superstructure is one that sits on top and is used to manage and guide your business. Discussion of a management superstructure is far less common than that of a supporting infrastructure, so I like to use an analogy in describing the importance of this.

Picture a warship at sea. The infrastructure of the ship comprises the ship's engines, the environmental support systems, weapon systems and the personnel required to efficiently crew and operate the ship. But these ships are not run from belowdecks where most of this infrastructure happens to be. The captain commands the ship from the bridge, overlooking the sea, the horizon and the ship itself. He is in direct communication with his crew, able to get a status report immediately at any point in time. And he has open communication lines to other ships and land-based information sources. From this vantage he gains a single unified view of the world in which he must operate, much as a CEO of a company requires to manage and guide his business.

This ship at sea also has other things in common with an operating business today. A ship at sea engaging in any military activity does not operate alone, much as no business today can operate in isolation. Instead, ships sail in fleets, much like businesses operate within integrated business communities. Just as operations of the ship must be coordinated with other members of the fleet and in concert with higher command, operation of a business must be coordinated throughout the supply chain to satisfy the requirements of a higher command — the customer. Ignoring high command in the military is a sure way of destroying a career.

Ignoring your customer can have the same effect on careers as well as entire enterprises.

But there is another factor at play here that makes this analogy so relevant today. Where do threats to a warship come from? Certainly there are threats from other warships. Fortunately, the captain of a ship knows well in advance of an attack from another warship. Even without the aid of daylight and clear visibility, technology prevents one battleship from surreptitiously sneaking up on another. Furthermore, there is some comfort in knowing that a similar war ship has the same or similar speed and maneuverability. No one expects a battleship to turn on a dime. Attacks can come from the air, but they too can be "seen" coming from a distance and warships carry appropriate weaponry and technology to defend against these smaller, more nimble threats. But perhaps the most devastating threat of all is the one that is the least expected, that which comes sight unseen, literally out of the blue. Submarines can literally and figuratively sneak up on a ship and launch torpedoes without being detected.

Where do competitive business threats come from today? Certainly every business has a traditional competitor — those enterprises in the same business. Typically these sources of competition are well known and, like a battleship in the distance, can be monitored. However, with the emergence of the dot-com companies in recent years it has been proved that competition can emerge overnight like warplanes entering the skies overhead. Much like these maneuverable aircraft, cumbersome or outdated business processes that might keep them from turning on a dime do not burden these companies, operating as virtual enterprises. But, do these companies have the necessary infrastructure to support growth and continue to provide adequate fulfillment services? The dot-com death watch that has occurred even more recently has proved that solid businesses, with established infrastructures, may take an occasional hit, but they can weather these threats. It is when competition springs from a totally unexpected source that it can be the most destructive. An example here might be helpful.

The branch of a major bank, call it Bank A, located in a thriving metropolitan area started to see its revenue from credit cards dip sharply. Upon investigation it became obvious that the demographic segment that represented the largest share of the drop was professional women in their 30s and 40s. To make a long story short, this was the demographic target of a new concierge service that had opened in the area. These women were generally upwardly mobile and led very busy lives both personally and professionally. This concierge service provided them the convenience of having simple errands and chores done for a fee. The new company had formed a partnership with another local bank, call this one Bank B, and in doing so offered a credit card, which, when used to pay for the service, offered a modest discount off the fee. So the branch of Bank A was torpedoed not by a direct competitor, not by another bank, but by an enterprise in another business entirely — a concierge service.

However, an interesting aftermath to this story came about a year after the concierge service company emerged. After a successful partnership with this competing Bank B, Bank B decided to terminate its relationship with the concierge service and a month later offered the same such services itself. So now the concierge service was itself torpedoed not by any known competition, but by its former partner — a bank.

BUILDING A MANAGEMENT SUPERSTRUCTURE

The management superstructure of a battleship is located on the bridge. It consists of the guidance and communication systems the captain needs to command the ship and provides a clear view of the horizon. The management superstructure of a business must have comparable guidance and communication systems and similarly must provide a unified, yet a broad view of the economic, financial and market horizon.

Starting with what would appear on the surface to be the simplest of these superstructure requirements, let's examine the executive's ability to view her business. Executives today are literally sitting on a goldmine of information residing in their transaction-based systems; yet still the information doesn't tell the whole story. Whereas the lament of management used to be information scarcity, the equilibrium among the production, the processing and the dissemination of information has shifted.

INFORMATION OVERLOAD

Enterprises find new and different ways to collect data every day, along with ways to automate the collection. Even information from external sources, about external influences is abundant. Never before has there been so much information available concerning world events.

The media constantly bombard us with business and information news. We see it on television, we see it in print but even more importantly we have access to it over the World Wide Web. We are literally awash in a flood of information and far more data than any one person can digest. Today executives are challenged to combine data collected from internal sources with that available from external sources, via the World Wide Web. The challenge then is to avoid information overload by intelligently selecting the information that is presented, and to present it in a way that is intuitively meaningful.

Leaders who are best able to gather business intelligence from all sources, and effectively utilize potentially enormous volumes of such data, are best able to affect, not only their bottom lines and therefore profits, but also their top line, increasing both revenue and market share.

Many CEOs still rely exclusively on subordinates to collect, process, summarize, translate and generally communicate information. The smart ones surround themselves with experts in finance, operations, marketing, engineering and any other area where domain expertise is required. These senior managers are hired not only to do a job but also to add inherent value to the decision-making process. And, of course, middle managers and knowledge workers, who assist them in turning data into information that must ultimately be turned into knowledge, support these senior managers.

Three factors are in play here that make relying exclusively on this human value-add process less than optimal. First of all, it takes time. Second, no one likes to bring bad news to the boss, especially if it makes the bearer of the news look bad. And, finally, even senior managers seldom see the full picture — that's the job of the guy or the gal at the top. Yet if information is delivered at different times of the month, or if it is filtered or embellished, a clear and unified view never emerges and it is enormously easy for that view to become fragmented and distorted.

Hence the emergence of EIS — Executive Information Systems. The first and foremost function of a traditional executive information system is to summarize. Yet more often than not the net effect of summarization is the masking of problems. And the higher the level of summarization, the more deeply problems and potential problems can be buried. Revenue is a key performance indicator that is universally monitored in enterprises across the world. If revenue is only 75% of goal, chances are each territory will be scrutinized. Yet if sales are 110% of quota, there is less chance of senior management being aware of territories that consistently underproduce. In turn, the net effect is lost opportunity.

Therefore, more traditional EIS are being replaced with what might be called a digital dashboard. You will find a rather broad range of specific business applications for just this purpose.

Before selecting such a solution, consider again our analogies of either driving a car or commanding a warship. How useful would a gas gauge be in your vehicle if it only told you how much gas you used yesterday? Or a navigational console that told you exactly where you were as of the end of the previous day? Sound silly? Of course, but how often do you attempt to navigate the course of your business with just such restrictions? How many CFOs or controllers know how much cash is available at any time during the day? How many sales executives know exactly where they stand in terms of goal, right now, in real time? Not where they stood yesterday, or worse yet, at the end of last month.

Therefore, whether your software provider calls it a digital dashboard, a command-and-control console or an EIS, consider if the information is derived and displayed in real time. The use of a data warehouse is an immediate signal that this is not the case. The mention of file transfers, downloads and sometimes even the creation of meta data are hints that what is displayed is not your most current and up-to-date view of your current business.

In addition, consider how that information arrives at your desk. Is it delivered manually in a report or is the delivery automated? Even if the delivery is automated, is it necessary for your CEO to ask for the information before it is delivered? In other words, does she have to pull it out of the system, or can she expect that appropriate alerts and notifications are pushed to her desk in real time? How useful would a temperature gauge in your car be if you had to ask it if the engine was overheating? Would it be acceptable if the warship commander had to periodically ask if there were hostile threats on the horizon? Is running your business any less critical?

ONE COMPANY'S APPROACH

Myers Industries of Akron, Ohio has taken a unique approach to using technology to provide its CEO, Steve Myers, with a clear picture of his business. Myers Industries is an international manufacturer of plastic and rubber products for industrial, agricultural, automotive, commercial and consumer markets. Myers manufactures more than 10,500 standard products ranging from plastic bulk containers and flower planters to rubber original-equipment-manufacturer (OEM) parts and tire repair patches. It also custom-molds products to fit customers' specific needs. Myers is also the largest wholesale distributor of tools, equipment and supplies for the tire and automotive underbody service industry in the United States. Myers is a public company, listed on the New York Stock Exchange under the symbol MYE. Sales in the year 2000 were $652.7 million.

Chances are you are familiar with one of their products:

- Reusable plastic material handling containers, tanks and pallets
- Plastic flower planters and trays
- Home storage and organization products
- Rubber OEM parts
- Tire repair and re-treading products
- Reflective highway marking tape

Myers is also the largest U.S. distributor of tire and automotive underbody service tools, equipment and supplies including:

- Tire valves and accessories
- Tire changing, balancing, and alignment equipment
- Service equipment and tools
- Tire repair/retread supplies

Myers Industries has grown by acquisition. Today it operates 25 manufacturing facilities in North America and Europe, 42 distribution locations in over 31 states, resulting in the operation of many different ERP and other related business information systems.

Because each operating unit or division of Myers is allowed a degree of autonomy, both in plant operations and information management, the result was a proliferation of business systems, which was preventing Steve Myers from gaining a consolidated and unified view of the business.

In addressing this issue, Steve's first thought was a conceptually simple solution that was a typical approach in the late 1990s, which was when Steve first sought a solution. Driven partially by the commonplace existence of legacy software during the time when Y2K loomed, the prevalent thoughts at the time were to replace heterogeneous mixes of business systems with a single, common ERP system. Major ERP vendors seeking to continue with the double-digit growth of the past decade further promoted this strategy. It was with these thoughts that Steve approached Andrew Winer, his CIO.

While the potential outcome of such a major overhaul of multiple systems was appealing to Steve, Andrew quickly determined the cost to Myers Industries. This cost was not only in terms of new software licenses, but also in terms of the effort involved — the blood, sweat and tears of which ERP implementations are made. Then he considered what the net result would have been. In many of the divisions it would have resulted in the implementation of different systems, but not necessarily better systems. In the end, these divisions would have expended time, effort and cash to get back to exactly where they were when they started.

As a result, Andrew and his staff considered several alternative approaches. Their first step was to re-define the problem. They found running multiple ERP systems was just one element. The real problem was business information. The data in the ERP systems were structured to present a transaction-based perspective of the business, not a business and management-based perspective. Yet presenting all the data and information that reflected this perspective would have created information overload. And while much of the data was contained in multiple ERP and other types of systems, there were also enormous amounts of completely unstructured data. Much of the information they had was in an unusable format, or they had no way of getting at it in a manner that presented a total corporate-wide, consolidated view. And the information to which they did have access was not timely enough.

After they ruled out new ERP implementations as too slow, too disruptive and too costly, they considered various subscription-based services and information providers, but they only exacerbated the potential problem of information overload. They considered specific software that would consolidate their financial information, but that left many other areas not addressed at all. What they found was that each of these solutions provided a piece of the puzzle, but none provided them with an overall solution, an overall picture of their business or a vision to build their future on.

ALIGNING BUSINESS STRATEGY WITH INFORMATION TECHNOLOGY

Having discarded these alternatives, Andrew then took a unique approach in coming up with a proposal. He went back to the root of the issue — the underlying business strategy. Myers Industries' overall strategy was growth by acquisition and internal consolidation in order to create a broader market presence in select markets. For Myers to achieve this goal, the executive management had determined it was necessary to accomplish the following:

- Quickly integrate operating units
- Reorganize internal structure quickly
- Be flexible in reacting to or creating market applications
- Sustain growth through internal and external synergy

Historically technology had been used tactically as an operational solution. But Andrew's new approach was to select technology that was strategically aligned with the business strategy. He asked the fundamental question, "What would software have to look like and how would it have to behave in order to mirror these objectives?" The answer was the following:

- It must integrate disparate systems quickly.
- It must limit the risk of changing mission-critical business applications.
- It must facilitate the deployment of new technology to users of these disparate systems globally.
- It must notify management when market conditions do or might change.
- It must support proactive management decisions.

What started out as a simple search for software had now taken on a new flavor. What Andrew found himself doing was attempting to overcome the classic disconnect between the senior executive and IT. Therefore, to address the immediate concern of providing his CEO with a unified view of the business, Andrew sought to build a technology-enabled superstructure on top of his existing business applications. He also sought an environment where new applications could be universally deployed, yet easily connected back to those disparate systems, sometimes replacing functionality that existed, sometimes expanding it.

Andrew sought an alternative approach to typical Enterprise Application Integration (EAI) solutions, which require specific interfaces to be established between multiple and specific applications, because this type of solution is not only costly, but difficult to maintain as new acquisitions occur or the underlying business applications change — not to mention the hassle of normal upgrades and maintenance releases for these systems. He found such a solution from interBiz, then a division of technology giant Computer Associates.* Several of the Myers divisions were already running one of interBiz's ERP systems, PRMS. The additional software they purchased and implemented was BizWorks, which interBiz refers to as an E-business Process Management Suite. While Computer Associates focused heavily on systems and development tools, interBiz focused primarily on business solutions that target the business processes that make up your business. The close relationship of the division, at the time, however, allowed interBiz to leverage leading-edge technology to these business processes.

Although BizWorks (graphically depicted in Figure 2.1) actually provides a broad base of functionality, the feature that specifically addressed Andrew's immediate requirements was its ability to provide a framework for shared data. Myers recognized that the basis for the consolidated view of their business was the sharing and the exchange of data managed by their business applications. The "wrapper" technology embedded in BizWorks was the key.

CONNECTING MYERS' DIVISIONS

What is "wrapper" technology and how does this work? This technology is based on the same principle as currency conversion with the introduction of the European Monetary Union (EMU) and the euro. When the euro was first introduced, of course it coexisted with individual country currencies. Countries within the EMU no longer performed conversion directly between countries and currencies. Values were converted first to a common currency. To convert from British sterling to French francs, pounds were converted to euros and then euros were converted to francs.

---

* The interBiz division of Computer Associates was later acquired by SSA Global Technologies.

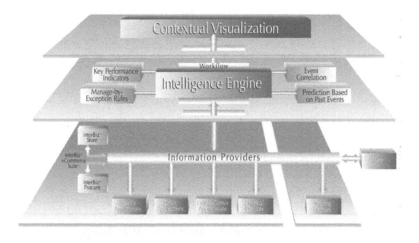

Figure 2.1   BizWorks: E-business process management. (Source: Computer Associates, 2002.)

In this way, global maintenance of conversion factors is simplified. When the value of the French franc changes, one single conversion factor is updated, instead of updating its relationship to all the currencies throughout the European countries.

Wrapper technology makes use of common objects in the same way as the EMU makes use of a common currency — the euro. Only the objects in this case are business objects such as customers and suppliers, along with associated data, purchase orders, sales orders, etc.

The wrapper technology, in this case BizWorks, allows Myers to potentially attach a "wrapper" to each business application across the corporation. The wrappers expose data in the business applications, through a common object model, containing business objects like purchase orders and sales orders, inventory values and accounts receivable balances, to BizWorks. All ERP systems manage these types of data. Yet they each store them in a unique way. In the past it was necessary for any application wishing to talk to another application to know how these data were structured and stored — just as it was necessary to convert directly from British sterling to French francs in the past. A specific conversion rate was required between the two, and between each of these and every other European currency. But wrapper technology simplifies this by serving as an interpreter. By exposing data through a common object model to BizWorks, it is open and available to any other application connected to BizWorks.

In this way, Myers eliminates the need for each application to "talk" directly to any other application. Instead they use BizWorks as a conduit, translator or interpreter among them all, and these applications become

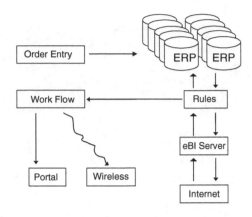

Figure 2.2   Dynamic customer risk assessment at order entry.

information providers to BizWorks and to Myers senior management. This allows them much more flexibility to plug in new information providers easily with each new acquisition, and to unplug any specific applications and plug in a different one as needs change and software is updated or replaced.

An example of how Myers made use of this technology is the implementation of what it calls a Corporate Enterprise Integrator (CEI). See Figure 2.2. One of the specific requirements Steve Myers had was to be able to assess quickly the relationship Myers Industries as a whole had with its customers. Because of the complementary nature of the products produced at the various divisions of the larger corporation it is not surprising to find that it was not only possible, but very likely that multiple divisions shared common customers. Each division therefore had a piece of the puzzle that told the whole story about credit history, buying profiles, as well as Myers' performance to that customer in terms of quality and schedule achievement. Yet because each implementation of the business system that would ultimately be the source of that information was relatively autonomous, these customers were known by different customer numbers or codes and the structure of this similar data could vary widely from division to division.

At this time Myers has connected 80,000 of the customers, with detail contained in five different divisions, through this CEI, and management can go to one source, one system, to view a consolidated credit, purchase and performance history of any one of these customers. And the company has done this without writing specific interfaces between any of the underlying systems. This type of approach to enterprise application integration — the use of this "wrapper" technology — specifically addresses the first three objectives Andrew sought to achieve in connecting Myers' business strategy with technology.

1. It integrated disparate systems quickly.
2. It limited the risk of changing mission-critical business applications.
3. It facilitated the deployment of new technology to users of these disparate systems globally.

Further detail on integrating information assets is provided in Chapter 6.

NOTIFYING MYERS' MANAGEMENT OF CHANGING MARKETS

The fourth characteristic of technology that Andrew and Myers Industries sought was the ability to notify management when market conditions do or may change. Obviously the path to this solution was not through the mining of data, which existed in the internal business applications. For this insight they needed to turn to sources of data elsewhere. Clearly, there is an overabundance of information regarding their markets available, but in order for technology to be useful in reaping the benefits of these sources it was necessary for that data to be available electronically. Naturally this led them to the Internet and the World Wide Web. This piece of the solution is depicted in Figure 2.2 as eBI (eBusiness Intelligence).

Myers prefers not to reveal what kind of data it retrieves in this exercise, beyond the capturing of customer and supplier data. Nor is it willing to publicly state what it does with that data — an obvious indication that Myers feels these company secrets have the potential of giving it a serious competitive advantage. However, it is very willing to talk about the technology used to capture and monitor this data. The key here is a unique combination of monitoring agent technology, combined with a search engine.

SEARCHING FOR HIDDEN FACTORS

While there may be many different technical approaches to search engines, from a lay point of view I find they generally fall in a line between a standard key word search at one end of the search spectrum, and full Natural Language Processing (NLP) at the other. Anyone who has used standard search engines such as Yahoo!, Lycos or Google is familiar with key word searches. You have a question? Key word search engines, which are free and available over the Internet, have enormous potential. Choose your favorite search engine, enter a few key words, and *voilà*! You have the possibility of getting more information than you could possibly have imagined. Some of it can be enormously useful. But it also has the potential of being totally useless and irrelevant. And wading through the junk in search of the gems is time-consuming and oftentimes unproductive. The more Web sites it finds and the more data it retrieves, the more you realize that you don't need more information to process. What you need is an answer to your question.

Unlike standard key word search techniques, NLP understands the structure in which a question is asked. Sequences of words create context and that context is critical in sorting through mountains of data to find a meaningful answer that provides value to you and your business. And common phrases can mean entirely different things in the context of different fields of expertise, industries and even across different geographies. Terms like "going south" mean something entirely different in the context of the leisure travel business than they do in the financial district. When several company names are mentioned in the discussion of bankruptcy, mergers and acquisitions, the structure and context of the text is important in determining who is acquiring whom, or which one is going bankrupt and which is acquiring the assets.

Standard key word searches represent a trial-and-error process, at best, because they have no way of knowing what is "correct." A good NLP, on the other hand, not only does a contextual search of the full text available, but also employs fuzzy logic — mathematical algorithms to allow uncertainty, or fuzziness — to be part of the search. As a result, it is then able to effectively rank the results to determine the potential value of the answers found; that is, how fully your question has been answered.

As an example, the first question I asked a search engine that used NLP was "Who is Cindy Jutras?" This was a search engine that employed a concept of ranking of 1 to 100%, indicating the probability that the information found was indeed the answer to your question. It came back with 28 possible responses, 27 of which had a 99% rating. Of those 27, 26 were me — references to speaking engagements, interviews, published articles, etc. The single irrelevant site with a 99% rating was an article, which referenced another Cindy Jutras who happened to be a teacher's aid someplace in the northern Midwest. Further qualifiers in my question could potentially have eliminated this red herring, but NLP and the process of posing questions using it was new to me. The 28th, which had a lower rank of 50%, was an article about a female basketball player — obviously no connection to me. Come to find out the search engine had encountered an occurrence of the first name "Cindy" and a last name of "Jutras." Both had been mentioned in the article, but not together, therefore the lower likelihood that the name referred to me. All in all, if someone had been researching Cindy Jutras, it would have taken less than 2 minutes to ask the question, collect 26 valid sources, and quickly eliminate the irrelevant data.

By analyzing the entire content of questions, pages and documents instead of key word approximations, users spend the minimum effort to get maximum results. Because NLP understands the structure of the question, it can distinguish between "what," "where," "why," "how many" and "how much," giving you an answer instead of simply pointing you in the general direction, or any number of directions, where you might find the information you need to draw your own conclusions. A good NLP will also incorporate rules that implement specific syntactic or semantic correlations in the context you specify. Rules are useful in separating out responses containing words such as "rumor," "allegation," "suspicion" and "doubt." Rules can tell the search engine that "going south" isn't a good thing, and reference to a company being "in the red" isn't making a fashion statement. The more specific the rules, the higher the precision of the results.

The obvious advantage of the standard key word search engine is that they are as immediately available as your nearest Internet connection. And they are free for anyone to use. The sophistication of NLP is available today, but obviously requires a software purchase. The trade-off between cost and value will depend upon the volatility of markets and economic conditions directly affecting your business, but the accelerating pace of both business today along with the pace of change itself makes the added value of NLP worth considering.

RECRUITING INTELLIGENCE "AGENTS"

As you may recall, the full statement of Myers' fourth requirement was to notify management when market conditions do or may change. A human being — making phone calls and searching through news sources either manually or electronically using the kind of search engine technology available at no cost could certainly have done all of the research noted above. But this relies first of all on that human having and taking the time to do the research. And what triggers the human to initiate the search? And what if there is no visible event to trigger it? Do you spend the time, all the time, on the singular chance that it will lead to some relevant findings that will justify the time spent searching and researching? During these times of leaner, flatter organizations in which we all are expected to do more with less time and fewer people, the ability to know *when* to instigate this research is essential.

Credit managers at Myers Industries are fully capable of performing credit analysis and determining credit limits. Their buyers are perfectly capable of evaluating potential suppliers in terms of financial solvency and other vendor selection criteria. The pace at which new customers and new suppliers are added is manageable. Yet potentially variable economic conditions don't wait for their periodic review cycles. However, the thought of performing these checks every time an order is placed would be enough to panic most credit and purchasing departments. And if nothing special has occurred to trigger a reevaluation, then it is not necessary. Yet again, in some instances, waiting for the next order may be too late. It may be too late to find another source for critical material. Revenue projections, which provide the fuel for growth, may be based on forecast predictions for a relatively small number of key customers. If business from any one of these key customers is in jeopardy, it can put the entire company at risk.

It was this latter risk that prompted Myers Industries to use agent technology to automate the process of searching for information regarding mergers, acquisitions, bankruptcies and other financial information for selected key customers. Once these agents were developed, they could be launched on demand, or scheduled to run periodically. So for these select customers, Myers does not wait for an event to happen, it simply constantly keeps an electronic ear to the ground for any news that might appear about these companies.

Myers treats supplier evaluations in much the same way. As a plastics manufacturer, many products are made from a relatively few number of raw materials. This makes the suppliers of those raw materials critical to the smooth operation of the manufacturing process. It makes sense then to keep relatively few market conditions under very careful and constant scrutiny.

Other companies may prefer an alternative approach. Where business is spread more uniformly across a broader range of customers or where there is a diversity of products and components, it may be more prudent to have the placement of an order trigger a specific search, notifying credit managers or buyers if a credit or vendor reevaluation is recommended. In this case an agent essentially "listens" for an event and may trigger another event. The event the agent listens for is the placement of the order. The event the agent triggers is the search for news items regarding the financial or operational status of the company placing the order.

Regardless of your application of this technology, it is important to understand the relative value this provides you. Employing these agents essentially provides you with an employee that is capable of working 24 hours a day, 7 days a week, 365 days a year. This is an employee who never gets tired, never gets sick, takes no breaks or holidays and never has to attend a child's soccer game or dance recital. This is an employee who works faster than a speeding bullet and never introduces his or her own prejudices to cloud issues.

MANAGING PROACTIVELY

Andrew Winer had determined one final criterion in selecting software that would align Myers Industries' business strategy with its IT:

The software must support proactive decisions.

The approach to monitoring selected key customers and suppliers on a regular and periodic basis was not only a means of notifying management of changing market conditions, but also an example of proactive decision making. However, in addition to the use of agent technology and search engines, there was one other element of technological innovation that factored heavily into the overall IT strategy — the use of predictive technology known as neural learning agents, or artificial intelligence.

USING EVENTS OF THE PAST TO PREDICT THE FUTURE

I have likened driving your business to driving a car, recognizing that leaders cannot drive their businesses based solely on internal information, any more than drivers can navigate based only on monitoring gauges and looking in the rearview mirror. Instead, drivers must also look ahead, collecting data from both internal and external sources, in order to anticipate conditions in the future. The safety of the passengers of the automobile is impacted enormously by the driver's ability not only to see the road ahead, *but also to predict the actions of other drivers and the results of those actions.*

It is becoming more and more critical for business leaders to focus on the future and predict business events as well. Conventional performance metrics have focused on the past, and at best might give an occasional, fleeting glimpse at present status. Pioneering executives today turn to predictive technologies, commonly referred to as artificial intelligence or neural agents, to complement and supplement their analysis capabilities. They employ these tools to anticipate scenarios and practice "what if" strategic management.

Neural learning agents are an example of predictive technology. They work much like the human mind, collecting data, observing conditions that are present when events occur, using that experience to learn and predict when similar events will recur. However, these neural agents have none of the limitations of the human brain in terms of the amount of data they can absorb, assimilate and analyze.

Artificial intelligence is not a new technology; yet it has traditionally been applied to specialized applications such as statistical process control within the chemical processing industry. Measurements of temperature, viscosity, specific gravity and other variables can be automatically collected and analyzed to recognize patterns and predict when a process is about to drift out of control. The same technology has also been applied to computer systems to monitor network, device or database performance to predict and, therefore, avoid system failures. It is now time for these same technologies to be applied in a business context.

If we can predict when a chemical process or a computer will fail, why not use the same methodologies to predict buying patterns of consumers, or market trends, attendance at a sporting event or currency and stock fluctuations? Companies that operate on a worldwide basis or seek to enter new markets are faced with a task of sorting through data whose volume has increased by multiple orders of magnitude. The sheer volume of data that must be analyzed has been enough to drive these companies to use standard statistical models, which result in only partial success — models that may ignore demographic differences and preferences in order to apply consistent rules to distribution or event planning. But neural learning agents, unlike the human brain, can comprehend and retain virtually unlimited volumes of data. They can recognize patterns that may not be obvious using traditional modeling tools. The application of predictive technologies is particularly useful in the transformation of data into useful knowledge. And after all, knowledge is power.

MYERS PREDICTS FAILURES BEFORE THEY OCCUR

With the specific objectives of producing higher-quality products, greater customer satisfaction and improved information access, Myers Industries' first milestone was to gain increased control over shop floor processes. It deployed a specific brand of neural learning agent technology, again from interBiz and Computer Associates, called Neugents. By employing Neugents in conjunction with BizWorks and its wrapper technology, Myers was able to connect this process back to its ERP system. It developed a Neugent to monitor plastic injection molders to predict when products may be manufactured outside of quality parameters, in order to reduce product scrap rates and improve on-time delivery and communication with customers. The flow of this process is shown in Figure 2.3.

Myers collects data from the injection molding process relative to temperature, material flow and other variables, and feeds that to a Neugent. The Neugent predicts when and where a molding machine will perform below standards, and because the process is integrated with the ERP system, PRMS, it is able to determine the impact of the potential failure in terms of which orders are impacted, and in turn, which customers. Myers not only can avoid producing poor-quality parts, but also can initiate

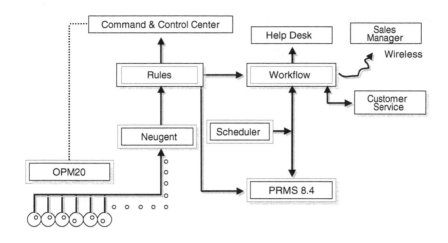

Figure 2.3   Predictive manufacturing models linked to key customer relations.

rescheduling, to minimize any negative impact on the customer and on productivity on the shop floor. The second phase of the process was to notify customer service representatives via alerts to their desktops and sales management via two-way pagers.

Therefore, instead of waiting for processes to drift out of control, or shutting down for preventative maintenance when it is not needed, Myers proactively monitors the situation and responds only when failure is predicted based on past experience. Predictive technologies are further discussed in Chapter 7.

SUMMARY

In this chapter we have first recognized technology as an integral component of business and discovered that business leaders today *must* understand technology and its enormous potential. While that may be bad news for some, the good news is that those who understand the opportunities that are derived from this technology have the potential of gaining a unique competitive advantage. Technology can truly be a means of competitive differentiation, particularly when it can be aligned with business strategy.

These C-level executives and senior managers will build an information superstructure, leveraging the vast amounts of data in their current business applications and extending their reach for data beyond. They will employ technology to gain a unified view of their disparate systems and from data mined from the Internet and other sources. To make efficient and productive use of staff they will take advantage of agents to constantly monitor these sources of data, as well as advanced search techniques such as NLP. And because their ability to manage change is directly proportional to their ability to anticipate it, they will look to predictive technologies to prepare themselves.

3

COMPANY CULTURE — ITS IMPACT ON YOUR FUTURE

How does company culture affect your ability to transform your business to meet the challenges of the world of E-business? E-transformation requires fundamental changes in the way you do business. How your company culture treats change, how accepting it is of change and the speed of assimilation are all significant factors in determining not only the success of this transition, but also your fundamental ability to compete today and tomorrow.

New ways of doing business require new ideas. Does your company culture encourage innovation or does it stifle new ideas? Effective communication is a prerequisite in our new connected world. How effectively people communicate will directly impact your ability to interoperate in a virtual enterprise. Does your corporate culture support open and honest communication and information sharing?

What is the impact of technology on this culture? Technology has the potential of opening up lines of communication from the top down, and throughout the entire organization. It can serve to create a viable means of two-way communication up and down chains of command, including dialogues with top executives. This can lead to productive information sharing or it can lead to unproductive "cover your tracks (and your tail)" tactics. It can bring people and departments closer together with shared knowledge, or drive them farther apart by replacing "face" time with electronic communication. Technology can keep managers connected to information 24 hours a day, enabling quick decision making but also invading their personal lives and disrupting family relationships.

Traditional implementations of ERP systems have typically created silos of information, producing departmental information that can easily be dispersed and manipulated. But the Inventory Manager may have taken a different snapshot in time from the Production Manager, and Finance is using month-end numbers. It is easy to see how everyone can be meeting the departmental goals, maximizing functional specialization to the detriment of the whole business.

ERP systems have succeeded in producing traditional financial measurements, but oftentimes these traditional metrics are not reflective of where you need to be taking the company today. The way you formally measure departments or individuals leads to rewards, praise and penalties and ultimately sets the stage for how they perform and also how they behave. And how people behave is both the cause and the effect of the corporate culture.

All these reasons combine to mean that technology and business systems alone cannot guarantee that you will be able to maximize E-business profits. Much of that success is determined by the culture into which you insert these solutions.

HOW DO YOU DEFINE CULTURE?

Corporate cultures are not defined by published mission statements or by documented corporate values. They are defined by how people think, speak and act. The culture of an organization is less likely to be created consciously, than unconsciously based on the values and behavior of the top management or the founders. Once created, some companies expend a great deal of effort and energy maintaining this culture. In these companies it is something employees are very aware of; in other companies the culture lurks largely unnoticed and ill defined in the background.

Southwest Airlines is probably as famous for its corporate culture as it is for being the only major airline in the United States that has been profitable each of the last 5 years. I have seen how its employees dress, I have watched them board a plane and I have heard stories about the nutty commentaries their pilots and flight attendants make on flights. I have never flown on the airline, yet I am very aware of its corporate culture. The airline hires people who have a good sense of humor and people who perform well as a team.

On the other hand, for 2 years I flew between 70 and 100 segments on American each year, and all I can tell you about its culture is that its employees don't have nearly as much fun as those of Southwest.

Corporate culture is a major component of the organizational infrastructure that supports the business and leads it in a set direction. Culture is often defined as "the way we do things around here." The fastest and best way to determine the culture of a company is to listen to the stories that circulate. Who is idolized? Who is ridiculed? What activities command respect? Which generate disdain? Regardless of how visible or invisible it is, no true leader of a corporation doubts the power of the corporate culture. How to leverage that culture, if it is productive, or how to change it, if it is counterproductive, represents one of the biggest leadership challenges today.

ABSORPTION OF CHANGE

What spurs people to action in your company? Is it a set of carefully planned steps leading toward well-articulated goals? Or is it the latest and biggest crisis? Is the atmosphere one of dynamic adjustments to change? Or is the routine set by the "way we've always done it"? While agility is a requirement to success today, there can be a fine line between being a dynamic and adaptive organization and one in which the rate of change can be disruptive. Maintaining the balance between stability and flexibility is a significant factor in successfully guiding a business through the turbulence of changing political and economic conditions in the midst of what can only be described as a technological revolution. Change for the sake of the business is good. Change for the sake of change is bad. Adapting to changing business conditions and climate is necessary. Making change simply because you can or because someone with a certain level of authority didn't think things through the first time is disruptive and unproductive.

Some people crave stability. Others thrive on chaos. Your corporate culture will determine which kind of employee succeeds in your company and whether or not the organization will be able to successfully transform itself into an effective E-business.

I worked for many years with a close friend and colleague, Gary Layton. Gary is a Vice President of Marketing for Computer Associates (CA). While we worked closely and we actually work very well together, when it comes to getting things done Gary and I have entirely different philosophies. In school I was the kid who had her paper done several days in advance. Gary was the kid doing his homework on the bus on the way to school in the morning. Each of us applies the same approach we used in school to our professional work. As a result I tend to inspire a little more confidence in co-workers. There is always a certain comfort level in knowing that I will get things done when they are required. Gary is a creative genius and the source of many truly inspired ideas. Yet, even though Gary usually does meet his deadlines, it is not without a degree of anxiety, if not downright panic at the final hour.

Because CA has grown largely by acquisition, the culture serves primarily as a filtering, weeding and molding process. Gary was hired directly by CA in late 1983, a few short months before I joined ASK Computer

Systems, which was a CA acquisition in 1994. Not surprisingly, Gary's philosophy blends better into the general culture of CA than mine did during my tenure there. There is such a tolerance and even expectation of change that doing things at the last minute is actually more efficient. Because Gary rarely does anything until the last possible moment, he effectively rules out revision and rework. I, on the other hand, spent more hours and days revamping and rewriting more work than I care to recall.

As a result, while it is contrary to my general nature and a physical impossibility for me to wait until the eleventh hour, I adapted. Eventually I learned to wait until at least the tenth. In fact, CA could actually benefit from more careful planning and less disruptive change; yet it was the culture that forced me to adapt, rather than the other way around.

Conversely, there are companies that are so resistant to change that the culture becomes an inhibitor to what should be a natural evolution of business. Over the years, as I have watched various companies grapple with the challenge of selecting and implementing ERP systems, I have been witness to more resistance to change than you might imagine. Considering that, for the last decade or more, most of these companies were embarking on projects to replace existing applications, one would think that change would be a natural result and a normal progression. Yet many simply wind up replacing one system with a more advanced, more feature-rich one, only to constrain the new system by changing it or customizing it, or simply limiting the extent of the implementation. The net result is that the new system behaves very similarly to, if not exactly the same as, the system they are replacing.

One company in particular, which shall remain nameless, was forced to replace its homegrown applications in 1999 because these applications were not Y2K compliant. This was a manufacturer of flow-control valves and therefore about 60% of its revenue and 40% of its product volume was custom-designed or custom-configured valves. Its inside sales representatives, along with the distributors, had always specified the configuration of the custom products by devising a customized, significant part number. In fact, this is quite typical of this particular industry. Specific digits of the customized part number represented different options for dimensions, metals and coatings. This process resulted in many significant errors and problems since there was no way for the system to verify or validate the configuration designated. These problems would typically not be uncovered until the late stages of manufacture when they prevented the product from being built as requested. Or the customers themselves would discover the problems.

In evaluating ERP systems to replace its existing homegrown systems, the short list of three vendors eventually was pared down to two. One of these systems had a powerful, rules-based configuration engine, which would have eliminated the need for the field personnel to construct the significant part numbers. In doing so, the company would have been able to apply sophisticated validation rules to ensure the custom product was buildable and would meet the customers' needs. The company chose not to purchase and implement this choice because this ERP system did not allow the entry of the manually constructed significant part number. Instead, it stepped the user through a series of questions to configure and validate the design. But that was not acceptable because it wasn't the way the company had always done things!

Instead, the company chose an ERP system, which did not have the powerful configurator, and it further customized the software to allow it to perform the exact same, equally inefficient and problematic process it had always performed.

In this case, the resistance to change, and the culture itself, was a direct reflection of the attitudes and receptivity to change of the president of the company. He gave a directive to the leader of the evaluation, and later the implementation team. The directive was that the new system must initially behave just like the old system because people could not handle too many changes at once. The company would constrain the new system initially, then later expand upon its capabilities. Of course, as yet that has not occurred.

The president was of the opinion that people fear change. People don't fear change. In fact, in many instances they welcome it. People fear uncertainty. They fear the unknown. And ultimately they want to do a good job. If anything, the employees of this company feared failing to continue to do as good a job with the new system as they had previously done with the old. In a culture where learning is encouraged and mistakes are tolerated, this fear would soon have dissipated, and progress would have been made.

Instead, the culture forced them into a standstill while the rest of the industry continued to march forward. Fortunately for this company few valve companies are innovative, visionary and progressive companies. So while I am sure there are some that could be described in this manner, there are enough that cannot be so described to allow this company to remain in business. In some industries this simply would not have been the case. In more progressive industries where technology solutions are aggressively pursued, this cultural position could very well have been a death sentence.

With respect to its ability to absorb change, this company is at the opposite end of the cultural spectrum from CA, where some of its employees sometimes joke that "CA" stands for "Change Always."

Chapter 5 delves more deeply into E-commerce as a step on the longer journey to full E-business integration and the changes this brings. Suffice to say at this point that as you shift your focus outward, rather than inward, you will have to evaluate many of your business processes and at least some of them will have to change. The abilities of your company and your employees to absorb change, and assimilate it to improve productivity will have a direct impact on your success. And nothing influences this absorption rate more than your corporate culture.

THE "NOT INVENTED HERE" SYNDROME

If you observe young children, you will see that humans are natural mimics. That is how infants first learn to walk and talk. And the natural tendency to copy someone they look up to doesn't dissipate once they have learned these basic skills. Little girls play dress-up with their mothers' clothes, jewelry, and, yes, even makeup. Little boys "help" their handyman fathers. Every parent who has ever fought with a child over what clothes the child wears understands the influence of the latest pop star. And there is nothing an adolescent fears more than being "different." So why are we, as adult businesspeople, so averse to observing what other people do and copying them? How did original thought and action become so elevated in our society? Why do so many people insist on reinventing the wheel?

From the time our child mimics first enter school, we start to build barriers against them borrowing, adapting and importing others' ideas. It is considered cheating, a serious transgression, to copy someone else's work. Looking at your neighbor's test paper is cause for suspension or even expulsion. Plagiarizing someone else's writing is banned. Nobody would argue that this is as it should be, but as a result, is it any wonder that after 12 to 16 or more years of education, borrowing ideas just doesn't sit well with us?

I recently came across the following quote:

> *Keep on the lookout for novel and interesting ideas that others have used successfully. Your idea has to be original only in its adaptation to the problem you're currently working on.*

Is this a revolutionary new thought from a recognized management guru? Hardly. These words are attributed to Thomas Edison.

Many companies invest in researching their competition. Yet more often than not this research is used in efforts to differentiate a company from its competition. Do leaders look to see what other companies are doing to succeed in order to emulate them? If not, is it because openly borrowing someone else's ideas runs against the grain of years of education? Or are they simply constrained by their egos, believing that by imitating another leader they acknowledge that someone else "does it better"? If so, then they are not able to distinguish between simple imitation and innovative imitation. And innovative imitation is an excellent path to growth and learning and a renewable source of competitive advantage.

As companies today become less vertically integrated, and instead turn to new and innovative collaborative relationships with other companies, the climate has never been better for innovative imitation. Leaders potentially have much more access and insight into other companies through these cooperative and collaborative arrangements than they ever had before. It is not necessary to imitate a company in an identical business. Sometimes it is far more productive to imitate one in a complementary business. Technology brings these companies together. And as technology is used to coordinate business processes across multiple companies, management can either set the example of innovative adaptation or it can resist ideas not generated within its own companies. As much as the overall corporate culture descends from the top, so does the willingness to accept new ideas "they" didn't invent. The pace of business today simply does not allow for the evolutionary trial and error that is the natural by-product of invention. The chance to learn from other companies' and other leaders' experience is simply too good an opportunity to pass up.

And so, I have my own simple bit of advice for those companies that suffer from the "not invented here" syndrome. Get over it. Acknowledge the source where appropriate, but steal ideas shamelessly. Encourage original and creative thinking but recognize innovative adaptation and creative imitation for the value it can bring in your organization.

THE ART OF COMMUNICATION

Corporate culture often dictates the nature of and the level of communication within a company. There are three types of communication that can exert an impact on your success or failure, not only in transitioning to E-business, but in all your daily operations.

The first type of communication is one that has a profound effect on the other two. How well does the leadership of a company communicate with the employees? People crave information. Remember that what many mistakenly view as fear of change is really fear of the unknown. And communication is the universal antidote for the paralysis that can easily be caused by this fear. How, how often and what senior management communicates set the tone for all communication throughout the organization, including the way every employee communicates with authority, at any level in the company. The guy or the gal at the top is frequently the last to know when something has gone wrong operationally. While no one wants to take bad news to the boss, where communication from the top is fast, frequent and filterless, at least the bad news gets there. In a company where communication from the top is slow, infrequent and guarded, employees will be reluctant to challenge what appear to be misdirected efforts, or even downright bad ideas. And in an age when the generation of innovative ideas is necessary to compete, reluctance to run any good idea up the flagpole can prove to be a travesty.

If management regularly applies the policy of open and honest communication, there is more of a likelihood of the second type of communication, interdepartmental interaction, being open, honest and productive. In this respect, even the most die-hard opponent to the idea that was "not invented here," has a tendency to revert to his or her childhood instincts to mimic behavior. Just as a child will seek to copy an adult, in an effort to gain approval, employees imitate the behavior of their superiors when it comes to communicating with those from another department. If upper management is guarded when it speaks of important matters such as marketing, product or growth strategy, major world or industry events, or the financial condition of the company itself, how can anyone expect lower levels of management or individual contributors to behave any differently?

And, finally, the tone of communication from management, as well as that which occurs interdepartmentally, will directly affect the communication with trading partners in general, but with customers in particular. And the ability to communicate within an integrated business community is essential to the seamless operation of the virtual enterprise.

How many companies do you know in which what is communicated to the customer and the "real story" can be widely different? I wish I could say that my experience with companies has proved this to be a rare instance. That is simply not the case. And yet when your customers sense this discrepancy, even when they only suspect they are not hearing the whole truth, there is a loss of trust, and with a loss of trust comes a loss of loyalty. And with loss of loyalty comes lost customers.

Now I'm not talking about a witness stand here where you tell the truth, the whole truth and nothing but the truth. Full disclosure of trade secrets and company strategy is not a necessary element of a solid business relationship based on trust. But if your customers cannot get a straight answer about the status of their orders, then you have a problem.

In many industries the initial purchase of a product is only the beginning of a longer-term relationship with a customer. Whether the basis of that continued relationship is built on further product sales or after-market service, or some combination, trust is an underlying factor, if not *the* underlying factor, which determines the longevity of the relationship and the continued revenue potential the customer brings. The higher the service content of this relationship, the more important the trust factor becomes. And the heavier the role of service, the more critical the element of communication.

These types of relationships are particularly evident in the software industry. Virtually every reader of this book will be a customer of a software supplier, either as a consumer or as a user of commercial software purchased by his or her company. For users of business software, companies buy a software package and typically a maintenance agreement for bug fixes and enhancements delivered in the form of upgrades and new releases.

The relationships between software vendors and customers are also some of the most contentious. The communication between a software company and its customers is complicated by the nature of the product itself. Unlike a hard good that you can hold in your hand, the definition of software is soft. The ability of a software product to meet the customers' needs is certainly dependent on the software manufacturer's ability to produce good clean and reliable software programs.

But it is equally dependent on customers' ability to identify, recognize and articulate their true needs. What kind of relationship do you think our previously described flow-control valve manufacturer wound up having with its software vendor? Do you think the ERP implementation ever satisfied its business requirements to the extent that the implementation could be deemed successful? I actually lost track of this implementation about a year ago when the valve manufacturer's parent company came in and replaced most of the management in the division. But prior to this, the division had experienced a period during which it effectively shut down production because it was not able to release orders. The division had thrown out the affiliate of the software manufacturer that had sold the product and had begun the customization and implementation. And it was working with the software manufacturer directly in a struggling attempt to back out many of the customizations it had originally written to make the new system look and behave like the old system.

Certainly this is an extreme example, but unfortunately it is not all that unique. And, it does serve to point out that the success of such an endeavor is dependent not only on the corporate culture, but also on the ability to communicate effectively and work cooperatively with trading partners — in this case the software vendor.

IS IT ART OR IS IT SCIENCE?

When discussing the communication throughout an integrated business community, we need to differentiate between the science of communication and the art of communication. For those relationships with vendors and customers that are based on the purchase and sale of hard goods, the science of communication is necessary and perhaps sufficient, particularly when we deal with standard products. That science is based on the electronic communication of demand and supply — forecasts, orders, acknowledgments, invoices and even payments. But for those soft products such as software and services, or in the case where custom products are collaboratively designed, the communication becomes much more of an art because supply and demand are less easily defined. Everyone has these types of vendor/customer relationships, although some companies to a greater extent than others. And it is when communication becomes an art that the corporate culture can significantly influence it. Our nameless valve manufacturer made no attempt to build a relationship with its software vendor with open and honest communication. It felt the purchase of software was not in the mainstream of its business and therefore was treated casually. What it failed to recognize was the potential impact this purchase and the resultant relationship could have on the business itself.

TECHNOLOGY AND YOUR CORPORATE CULTURE

By far the largest impact technology has on the corporate culture results from its ability to open up lines of communication throughout the organization and beyond. The ability for a CEO to have direct communication with each and every employee of a company is very real today. E-mail and teleconferencing break down the geographical and hierarchical barriers and can potentially connect the highest levels of management directly with employees worldwide. Where fear of the unknown can cause delays and disruption, fast, filterless and frequent communication can be an effective antidote providing these means of communication do not wind up replacing face-to-face time where it is necessary or appropriate.

The potential for opening up a dialogue of two-way communication between employees and senior levels of management carries enormous benefits. It not only provides additional insights into concerns of employees, reinforcing a culture that supports open and honest communication, it also has the potential of uncovering opportunities for organizational growth and possibilities. But possibly the most important benefit is that employees feel more intimately involved in the business, and that ensures that they care, not only about their job and their contribution, but also about the overall success of the corporation.

Because of the possibility of unleashing an unprecedented awareness of issues, the very foundation of the corporate culture can either be reinforced or severely shaken. In an organization where corporate hierarchy is sacrosanct, this potential for direct communication can have an unnerving effect on members of middle management who previously perceived their role, and in essence their reason for being, as a conduit for the controlled dissemination of information. If the boss communicates directly to everyone, the individual managers may feel they don't add value in the chain. It can be enormously threatening to those who previously saw direct, or even indirect, communication with upper management as a measure of importance.

Yet for larger and even mid-size companies, the establishment of this two-way communication can be the equivalent of opening the proverbial floodgates. In companies of thousands or even tens of thousands of employees this can quickly get out of control unless there is a mechanism in place to make sure questions and concerns are answered without requiring inordinate amounts of personal time of top management. In these cases senior management needs to be aware of and willing to tap into further use of technology to assign surrogates for response that allow selective monitoring of responses.

ACCELERATING THE SPEED OF BUSINESS

As companies become more global, spanning countries and continents, communication for timely decisions becomes more complicated. Whether you are negotiating important contacts and contracts or simply staying in touch with employees, nothing is more effective in building relationships than "face time," time spent communicating in person, face to face. A natural consequence of this is the need for leaders to be more mobile. Yet these same leaders are required to make timely decisions to facilitate the necessary flow of business processes. Whether it is approving purchase decisions or special terms for a major incoming order, the speed at which business can be conducted is dependent on the accessibility of these decision makers regardless of whether they are sitting in their offices or traveling half way around the world.

This situation is further complicated by the global nature of business today. The more geographically dispersed an enterprise becomes, the less downtime there is when decisions are not required. Corporate culture will significantly affect how decisions are made and the levels of approvals required. The more centralized the decision-making and approval process, the more necessary it becomes for senior management to be consistently, if not constantly available. The nature of the decisions should dictate the urgency of the contact, but wired and wireless devices make it possible for managers to be in contact virtually 24 × 7. I say virtually because these devices can be shut off, and I, for one, am perfectly capable of sleeping through a page, even when they are not off. While the device can physically be disabled, culture and attitudes play a strong role in determining if it will be. Corporate culture can put enormous pressure on middle and upper management to be in constant contact, making the preservation of privacy and personal time impossible. In some corporations, elevation through the ranks of management means the development of buffers to more senior managers, making them more and more inaccessible to interruption. In other cultures, it is the opposite. The acceptance of more responsibility is accompanied by additional sacrifice of personal and protected time. Typically this doesn't happen by design, but is the consequence of one of two environments. It can be an indication of the most senior levels of management's disregard for the personal welfare of their subordinates, or indicative of the manager's desire to be perceived as both busy and important. For those with heightened perception of self-importance, this can be viewed and displayed as a badge of honor. For those simply driven to it by the culture, it can wear away their loyalty and, ultimately, it can have a detrimental effect on employee morale and therefore turnover. Yet even if these managers wind up quitting, it is usually not before they have wreaked havoc on their personal lives and relationships. While dedication from employees is a sign of the positive effect culture can have, unless a balance is maintained, the scales can be easily tipped in the wrong direction.

At first glance, technology appears to be a culprit in this scenario. After all, it is the wired and wireless communication devices that make this constant contact possible. However, it can also alleviate the problem by providing true management by exception through automated application of business rules and policies, coupled with automated notification and escalation of alerts. But as you investigate these enabling technologies you need to recognize that your corporate culture will pave the way for such technology to succeed or to fail.

SILOS OF INFORMATION

Prior to ERP, the development and implementation of individual business applications naturally created silos of information. Inventory Management, Order Processing, Engineering, General Accounting — each department had its own data to collect and transactions to manage. Batch interfaces did little except to preserve these departmental divisions. The introduction of ERP systems was intended to destroy these silos of information and provide a single, integrated system. Yet ERP systems were specifically designed around individual departmental needs. Yes, they employed common databases and they made the interfaces, if not real time, at least faster and more streamlined. Master and transactional data indeed was shared, but did this serve to break down the barriers between departments?

An ERP project typically starts off with the creation of a cross-functional team so that the interests of all departments are preserved during the implementation process. But once the master data is defined, how much interaction is preserved? While data may be shared in a common database, and this can be a major hurdle for many companies, the natural inclination for departmental managers is still to focus on their departmental requirements. And those departmental requirements will largely be determined by how each of those departments is measured.

ERP systems have primarily been designed to report traditional financial measurements. As a result, departmental managers tend to focus on the output of the ERP systems that signal to them indirectly or directly measurements of their performance. If the purchasing department is measured based on its ability to obtain the best unit price, then that will be its focus. If the inventory manager's goal is zero inventories, he will do everything in his power to initiate just-in-time inventory management and control. If the method by which the purchasing department drives unit price down is to increase lot sizes and purchase quantities, we have a conflict of interest. The ERP system alone does nothing to alleviate this conflict other than report costs and inventory levels. The purchasing manager will report to her superiors her achievements with output from the ERP system noting average unit costs. The inventory manager must then justify the rising inventory levels, also noted by ERP output.

ERP is able to provide a variety of measurements as a report card. How well did the purchasing manager do in reducing unit costs? What are current and historical inventory levels? But how do you analyze the interdependencies between inventory levels and customer satisfaction? Customer satisfaction is certainly contingent on delivery performance, which can be evaluated by comparing customer requested and actual ship dates. But what is the trade-off between price and delivery? How much more could you charge for your product to ensure delivery performance meets customer expectations or exceeds your competitors' service levels? And is there a correlation between higher component inventory levels and your ability to meet customer expectations in shipping finished products? And, finally, how important is delivery performance in overall customer satisfaction? Is it the number one factor, or does it slip to somewhere around third or fourth place? Not only are these cross-functional and subjective questions, but much of the input you need to consider, such as customer satisfaction levels and customer priorities, simply does not reside in the database of your ERP system. And, therefore, it is no wonder that the implementation of ERP was not the silver bullet that would destroy individual silos of information.

So if ERP is not the answer to breaking down the silos of information, what is? The answer is actually multidimensional. First of all, you have to remember the old adage, "what gets measured, gets managed." The metrics themselves are not the goal, but a means to an end.

Second, you need to reflect on whether the performance metrics available using a standard ERP approach provide sufficient tools to accomplish your strategic objectives. Strategic objectives are not departmentally specific, but reach broadly and deeply across multiple facets of any company. Plus you have to ask yourself if the performance metrics in your company have become the goal. Does the purchasing manager see the metric of unit cost as a goal in of itself, or does she recognize it as a means to an end — the containment of costs and profit improvement? The answer to this question will determine whether inflation of purchase quantities beyond what is reasonable and required will be the counterproductive result. The implementation of automated business rules and policies could provide a technological solution to this problem, but in reality this is a management issue and a people problem. And the level to which people understand and care how their actions affect the overall performance of the organization is a reflection of the attitudes and culture within the enterprise.

SUMMARY

At the beginning of this chapter I asked the question, "How does company culture affect your ability to transform your business to meet the challenges of the world of E-business?" The way people think and behave, which is the manifestation of any corporate culture, can either enable that transition or obstruct it. Your company's ability to embrace change, without letting it overrun the business, has an enormous impact, as does its ability to openly and honestly communicate both internally and externally. Technology can serve to open up lines of communication as well as provide performance metrics that go beyond the traditional financial measurements found in most ERP systems today. Those companies that suffer from the "not invented here" syndrome, that stifle creativity and that spend more time covering their backs than they do covering the bases will suffer. Those companies that have the ability to embrace change, while effectively managing it, that have the ability to communicate openly and honestly both throughout their organizations and beyond and that make innovative use of performance measurement as an effective management tool will be able to harness the energy created by their corporate culture.

4

THE EVOLUTION OF E-BUSINESS — WHERE DO YOU STAND?

Within the last several years, many industry analysts and business pundits have predicted that E-business will become so pervasive as to cause the "e" to be dropped and E-business will simply become "business as usual." Although this prediction has not yet been realized, we have reached an interesting juncture in the evolutionary process because the term itself, *E-business, has* become pervasive. However, in spite of the very common use of the term, the understanding and definition of the term is far from universal. To some, E-business is synonymous with E-commerce. To others, it is all about the World Wide Web. And to some it remains a distant illusion. If our goal is to understand how to go beyond ERP to maximize E-business profits, it only makes sense to understand what we mean by this term, and the implications it has for your business.

E-business is not a single state of being. More than one expert has referred to different levels or generations of E-business, and many times these levels or generations are presented in sequence as if the process of becoming an E-business is an evolution. This chapter explores this evolutionary concept of E-business in terms of business processes. It will describe three stages of E-business: E-information, E-commerce and full E-business integration. And while E-information usually is a prelude to E-commerce, and both of these are certainly prerequisites to full E-business integration, the path from start to finish may not be a strict linear sequence. E-transformation is a complex process because it requires people, processes and technology and it must be coordinated with strategy, business plans, tactics and operational policies. Depending on where you stand

1st Generation	2nd Generation	3rd Generation
E-Information	E-Commerce	Full E-Business Integration
Supports Existing Business Models and Processes	Streamlines & Automates Existing Business Models and Processes	Re-architects Your Business Model & Re-engineers Business Processes

Figure 4.1   The maturing of E-business.

today, there may be multiple, parallel and intersecting paths you might take to your final destination.

THE THREE GENERATIONS OF E-BUSINESS

While sometimes referred to as levels or stages, I have chosen to refer to these three as "generations" for a very specific reason. The generations of E-business have much in common with the generations of a family. While it takes a certain amount of time for a family to span three generations, once this level of maturity is reached, the growth process continues. At some point, of course, an additional new generation will be born, but in the meantime all of the generations, not just the most recent one, continue to grow and mature.

Figure 4.1 depicts these generations as an evolutionary process with E-information as the first generation, E-commerce as the second and E-business, shorthand for full E-business integration, as the third generation. Yet a company need not "complete" a generation before it moves on to address elements of the next. Nor does the preceding generation stop maturing when the next is spawned. If this concept seems a little fuzzy now, I hope it will become much clearer as we define the various terms.

We spend significantly more time discussing the first generation of E-information. Chapter 5 deals in depth and exclusively with the second generation of E-commerce and from there on we further flesh out the requirements of the final goal, full E-business integration.

BUT FIRST, SOME DEFINITIONS

Before we proceed, several definitions bear noting. For those readers with a more technical background, this may seem terribly elementary, but we all need to recognize a lot of terms are used today with little understanding of what they mean. So, for those who are well versed in this area, bear with me for a moment, while some of your fellow readers catch up to you.

First of all, an **Internet** is simply a global network connecting potentially millions of computers. It is decentralized by design. Each computer connecting to the Internet is called a **host** and runs independently from all the other computers connected by the network. Since most likely you will be connecting to the Internet from a personal computer (PC), however small or however powerful it may be, it too is a host and therefore other computers accessing the network can potentially have access to information stored on your PC. It is for this reason that the very technology that provides connectivity to the world through the Internet also puts us at risk for security.

Servers are computers that provide services over a network to multiple users. The services of this sort with which you may be the most familiar are e-mail services. Messages are delivered first to the e-mail server. As you connect to this server, you have the option of accessing these messages over the network, directly from the e-mail server, or more likely you will download these messages from the server to your own PC.

The **World Wide Web** (WWW) is a system of Internet servers that support documents specially formatted in **HTML** (HyperText Markup Language). These documents are special in that they support links to other documents and applications. Jumping to the other document is as simple as clicking on a hot spot on your screen, typically a graphic or a highlighted piece of text. To reach, or hyperlink to these other documents, a **url** (uniform resource locator) is used. This is a global address used to uniquely locate resources on the WWW. Although all Internet servers are not part of the WWW, for general purposes, the Internet has become synonymous with the WWW.

At the risk of appearing to use circular definitions, the definition of a **network** is a group of two or more computers that are linked together by some form of **communication protocol**. In this case the term *protocol* can be intuitively understood. As much as protocols between nations determine the way visiting dignitaries are treated, communication protocols determine how one computer deals with another. **TCP/IP** (Transmission Control Protocol/Internet Protocol) is the *de facto* standard for transmitting data over networks used by the Internet. It is simply a suite of communication protocols used to connect hosts on the Internet. Generally, knowing whether a network uses TCP/IP or not is about all you need to know, at least in understanding the concepts in this book.

Some networks are private and some are public, and some are somewhere in between. While an Internet is accessible to the general public, an **Intranet** is a network based on TCP/IP protocols, and therefore behaves just like the Internet; yet it typically is a private network where access is limited to employees of a company or members of an organization. Not only is access to an Intranet typically controlled by the identification of users by user names and passwords, it is typically sur-rounded by a **firewall** to prevent unauthorized **access**. A firewall is implemented with the combination of both hardware and software. While an Intranet sits behind a firewall and is accessible only to employees or members of an organization, an **Extranet** provides partial accessibility to outsiders, again through the use of user names and passwords.

There are many other terms related to Information Technology (IT), and we will define others as needed as we go along. However, this should suffice in preparing the reader for understanding the three generations of E-business that we describe in this chapter and throughout the remainder of this book.

E-INFORMATION — THE FIRST GENERATION

I conducted my own personal and somewhat random survey of acquain-tances and asked each a question, "What first comes to mind when I say E-information?" As a result of my relatively unscientific methods I found two answers predominated, each occurring with similar frequency. About half answered with "e-mail," and the other half replied with "having a company Web site."

E-information, put simply, is the process of sharing information elec-tronically. There is certainly no doubt that e-mail is an important and effective method of transmission for E-information, but e-mail itself should not be confused with the content, or information, it transmits. A company Web site is both. It is defined by its content and presentation and it also serves as a delivery vehicle for the information contained therein.

Intuitively, many people equate E-information to the creation of a company Web site, and in fact without one I would wager to say that an enterprise would never even reach the toddler stage of E-information. The establishment of a Web site has become quite pervasive. Figure 4.2 shows the results of a survey conducted as early as 1999 by AMR Research, a Boston-based research firm. A total of 800 companies, ranging in size from $30 million to $80 billion in revenue, were surveyed. At that time, over 90% of these companies, some of them relatively small, had estab-lished a Web site. The results of this survey confirm that the vast majority of companies had at least reached the early stages of the first generation of E-business, known as E-information. But how far had they matured in this generation?

The establishment of a company Web site represents only the tip of the iceberg in terms of the information that might be electronically shared. The audience to which the information is directed might best categorize this information. Table 4.1 lists different audiences. The most generalized audience is the public seeking product or company information, and

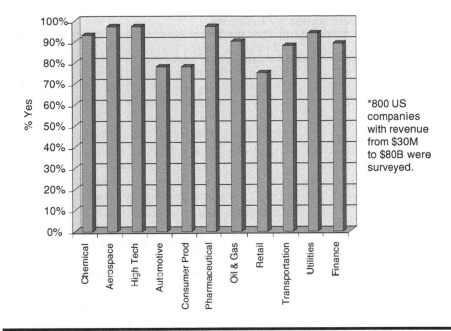

Figure 4.2 Numbers bear it out: does your company have a Web site? Results of survey of 800 U.S. companies with revenue from $30 million to $80 billion. (Courtesy of AMR Research.)

indeed a company Web site is certainly one way to extend the marketing reach of an enterprise. And just about every business today has one.

Yet the extent to which these companies use the Internet and their Web sites to provide E-information varies widely. At one end of the spectrum are the companies that use the Internet as an electronic yellow pages advertisement, with little else to be gleaned from the site. On the other end of the spectrum are those companies that rely almost exclusively on the Internet as the source of all marketing materials. Some companies, for example, no longer produce printed versions of their marketing brochures. Their Web site on the Internet is the sole source of this collateral.

I actually have a very simple benchmark for evaluating a company's Web site. I simply compare it to the Web site of my nephew's rock band. I can log onto www/dubbedover.com and find out just about anything I would ever want to know about the band. First of all, the Web address is intuitive. It is the band's name (Dubbed Over) followed by ".com." From the site I can read about the history of how the band was started, including an evolution of its name and the changing of band members over time. I know where they have played recently and where they are scheduled to play. I can find out how to get their latest CD and learn about others scheduled to be out soon. I can see what the band members

Table 4.1   E-Information by Category

Who?	How?	What?
General public (industry/marketplace)	Internet	Company Web site Electronic brochure Product descriptions Press center Directories and listings
Customers/distributors	Extranet	Online product catalogs Electronic product specifications/information Anticipated product demand (pull)
Suppliers	Extranet	Online product catalogs Electronic product specifications/information Product/supplier research Anticipated product demand (push)
Employees	Intranet	E-mail Bulletin boards Online manuals Job postings Organization charts Address/phone listings Performance metrics

look like, and I can hear a sample of their music and sign their guest book. And as I sign the guest book, I can vote for my favorite "Dubby" (band member), and optionally ask that the band contact me. I can also sign up to be notified of late-breaking band news and upcoming events. And by the way, my nephew is 17 years old.

So where do you stand? How does your Web site stack up against the Web site of four teenagers with no professional marketing experience? Let's take it feature by feature:

The Web Address Is Intuitive: How easy are you to find? There is increasing contention for truly intuitive Web site addresses or urls. As a result many companies secure several prior to settling on the one they feel will be the most productive. As a result, site names that were previously reserved can often be secured through persistence and a willingness to buy the rights to use them.

"About Us" — The History of the Band: Can I find out about your company? How was it founded and how long has it been around? Does your Web site establish your credibility in your industry?

Identity — The Evolution of the Name: Has your company always been known by its current name? With the recent merger and acquisition mania surrounding many industries today, this can be a troubling consequence. Are you leveraging all the brand equity or name recognition that you can?

Band Members, Like Board Members, Change over Time: Who are the key executive players in your organization? Has there been continuity and stability or a revolving door? If I don't see evidence of the former, I start to look for other sources of information, and these sources may not always be flattering or objective.

Where They Play — Product Endorsement: If I wanted to hire the band, what better way to check them out than to see them play in person at some scheduled, upcoming event? If that is not possible, perhaps I can talk to someone for whom they have played. Do you have showcase customers featured prominently on your Web site? Can I see how and where your products are used or the results of your services rendered?

Product Availability — Where I Can Buy Their CDs: No, the band doesn't have an electronic storefront, but then again the four are not quite a big enough sensation yet to accept credit cards. Are you? But I can send them an e-mail, call them on the phone or leave them a message in their guest book. And I guarantee that if you contact them in any of these ways, the CD can be on its way the same day payment is arranged. Do you provide multiple ways of contacting your company? How good is your fulfillment rate?

Product Visualization — I Can See What the Band Members Look Like: Obviously the interpretation of this feature will vary between industries and products. In Dubbed Over's case, the members of the band are as much a "product" as their music is. And for their target market of teenagers, how they look and the image they project is sometimes more important than the way they sound. Can visitors to your Web site see what your products look like? If you have a service-oriented business, the people delivering the service are equally important as the services they provide, and while how they look may not be important, what credentials they hold are.

Product Demonstration — I Can Hear a Sample of Their Music: Can visitors see or hear product presentations, hear customer testimonials or see your product demonstrated?

Contact Us — I Have Multiple Ways of Contacting Them: How does the visitor to your Web site contact you? On www.dubbedover.com I can get the cell phone numbers for two of the band members. Of course, when they are in school you must leave a voice-mail message. I can also get e-mail addresses for all four of the band members, or I can sign their guest book (either publicly or privately) and ask that they contact me by phone or e-mail. Do you make it easy for interested parties to contact you?

Sign Up — I Can Join Their "Notify Me" List: Do you have some way of bringing visitors back to your Web site, or a means of offering them information that keeps you and your name in front of them? While this is a valid and desirable objective, consider carefully how you go about this. Dubbed Over's notifications are only sent to those who confirm they want to be on the list *and* the band provides a convenient way to un-subscribe. Nothing irritates me more than innocently visiting a Web site only to have the visit result in a flood of unsolicited messages, each of which thanks me for subscribing and assures me messages are never sent unsolicited. Automatically signing up an unsuspecting visitor steps over the line separating aggressive marketing from unethical business practices. However, allowing visitors to subscribe to useful information is a service you can provide without abusing their privacy. I find a telemarketing follow-up phone call rather annoying. This is usually the result of request-ing specific "free" information such as white papers and research results. While an annoyance, I can rationalize this as the price I pay for obtaining the information I seek.

So, how did you stack up against my 17-year-old nephew's site? If indeed you visit this site, chances are you won't be particularly impressed by its look and feel. Yet that is perfectly understandable considering the audience a teenage rock band would be trying to reach is substantially different from my target audience as a writer.

While not the sole delivery vehicle of general E-information, your Web site is probably the most effective one. Yet, unlike Kevin Costner's *Field of Dreams*, you cannot be guaranteed that if you build it, they will come. A Web site is one of those unusual marketing vehicles that itself must be marketed.

There are numerous ways to do this, including traditional marketing vehicles such as directory listings, direct mail, public relations and adver-tising campaigns. Each of these traditional methods has a parallel electronic delivery vehicle. Many directory listings have gone online as a means to reduce the cost of updates and more effectively push these updates out to subscribers. Your ability to participate is gated only by the use of simple desktop tools and an Internet connection, and in fact the publisher of the directory will most likely control your migration to this media. The electronic equivalent of direct mail campaigns is generally viewed as spam, or junk e-mail. Unsolicited messages have about the same likelihood of generating real business as does the paper junk mail we throw away without opening, unless it has truly been solicited as with Dubbed Over's Notify Me List.

Of course, the electronic counterpart to advertising comes in the form of "banner" ads, those small, usually animated boxes of information that we see when we go to well-known sites such as Yahoo!, Google and cnn.com. Obviously there are numerous media for advertising, including print, television, radio and other assorted delivery mechanisms. Probably the most direct comparison of these "banner" ads is to print ads in newspapers and magazines. While most of us don't linger long on advertisements in newspapers and magazines, we can realize the full impact of a well-done advertisement as we leaf casually through the publication. This is not the case with a banner, which we need to "click" on to see the advertisement. Therefore the banner itself, usually quite small, must be compelling enough to spur you to action.

This new media for advertisement has one very specific advantage over traditional print and other media, however, and that is the ability to better analyze the response to the ad. Publishers can tell you the number of subscribers to their magazine, and you yourself can count the actual responses to the ad either via a phone call, a return "bingo" card, or some other response vehicle. But no publisher can tell you the number of readers who actually pause and read your ad. This is where technology can provide you far more insightful data including the number of visitors to your Web site from the banner ads, how long they stay and which pages of your Web site were visited. This is particularly valuable when your primary objective is building brand awareness as opposed to the purpose of generating specific leads.

Approaches to gathering this information can range from the purchase of a simple Web traffic reporting tool, to be installed and administered by your own Web master, to more expensive and sophisticated products or services, most likely provided by an advertisement agency specializing in new media.

These tools can tell you who is visiting, how they got there (e.g., from a banner ad, an online directory listing, a search engine or some other means) and how long they stayed. They can determine the typical paths that lead to purchase or inquiry. This can be particularly useful in determining how promotions affect the volume and quality of visitors and whether these result in actual orders.

The ultimate question in the mind of an executive approving the budget for a marketing campaign is, "How much business is the direct result of this promotion?" This is true whether the campaign is a simple banner ad or a more extensive traditional or electronic vehicle. To answer this question you will need to close the loop and integrate the results of the Web traffic reporting with other business applications.

When the customer is a consumer shopping online, there is a fairly quick and direct route to the business application that manages your sales orders. But in the case where a direct or indirect sales force is involved, this may be a more circuitous path to an order entry application, possibly by way of a Sales Force Automation (SFA) or Lead Tracking system. It is through the integration of these Web tools to your back-office applications that their value can be maximized. We talk more about application integration in Chapter 6.

COMMUNICATING WITH YOUR EMPLOYEES

So far we have been talking about the Internet and providing E-information to the general public. Let us now approach the more select audiences previously outlined in Table 4.1, starting with the audience within your company, your employees. For many companies, the birth of this type of E-information came with the introduction of e-mail, typically supported by a private network, which was accessible only to employees. Along with e-mail, or shortly thereafter, came electronic bulletin boards, which in many instances were introduced as a feature of e-mail, or at the very least also required access to the private network. If all employees were located in a single facility, a local area network (LAN) was sufficient to connect them. If the organization was distributed across multiple facilities, a wide area network (WAN) was required, supported by dedicated communication links. While e-mail systems became quite pervasive as early as the early 1980s, it was not until the mid-1990s that e-mail access expanded to include Internet access, with all of its myriad uses.

As we noted in Chapter 3, the possibilities for fast, frequent and filterless communication throughout all levels of a corporation can make an enormous potential impact on a corporate culture. And, conversely, the corporate culture itself will have enormous impact on the degree to which information is indeed shared. Regardless of these factors, the introduction of Internet access in general and Intranets in particular opens up all sorts of possibilities in the electronic sharing of information with your employees.

Let's look first at the concept of e-mail and the electronic bulletin board and the possibilities these brings for keeping employees informed of policies, procedures and internal company news. As I mentioned in Chapter 3, I came to Computer Associates by way of the acquisition of The ASK Group in 1994. ASK had standardized on HP mail and used this standard to connect all employees throughout the world by means of a WAN. At the time of the acquisition, and for the prior 10+ years I spent with ASK, this was an important means of communication, but it connected only employees internally. I didn't use a PC to connect, just a dumb terminal. In fact, until the year of the acquisition, 1994, I didn't even have a PC. When I came to Computer Associates, the company was using its own homegrown e-mail system, which at first I found difficult to navigate given its mainframe look, feel and approach, which was entirely foreign to me given the ASK orientation toward the minicomputer. It too was used only to connect to other employees internally, but even those few short years ago, that was pretty much the standard in corporate America at the time. One advantage I immediately had over the more intuitive HP mail was an elaborate electronic bulletin board, organized by topic, departmental functions and by geography. I could see notices that applied only to the Massachusetts offices, or those that were applicable companywide. There were travel and expense policies, and the bulletin boards were the means by which changes in policies were communicated. There were reminders and announcements about benefits and reminders if you had to re-enroll in a program like a 401K or estimates in a health care account. Of course at the time, these reminders were just that. We still had paper forms for insurance claims, 401K deductions, and had there been an employee stock purchase plan at the time, as there is today, that too would have required paper forms for the authorization of payroll deductions.

There were features that were not available that I take for granted today. For example, we were unable to send files using e-mail. The fax machine and I became great friends (actually it was more of a love/hate relationship), I had a drawer full of 3$\frac{1}{2}$-inch floppy disks (make that two drawers full) and I was a regular customer of Airborne Express and the U.S. Postal Service.

About 2 years later, in 1996, Computer Associates began phasing in Internet connections, Microsoft Exchange for e-mail and the implementation of a corporate Intranet. All of a sudden I was able to send files and access data published on the Intranet. I realized just how much better my life had become one day when I couldn't find the fax machine in the office. Come to find out, it had been moved to a different location 2 weeks earlier. And I hadn't even noticed! And I would love to have received the money I did *not* spend with Airborne Express in my paycheck as a bonus every month. Today every employee in the company has access to the Internet and the e-mail standard is Microsoft Outlook. Along with the ability to send files, I also have the ability to schedule meetings, manage my calendar electronically and receive those all-important reminders that signal me when a meeting or a deadline is approaching. My favorite feature has become the ability to preview messages. The first three lines are displayed in my Inbox, as my mail is being downloaded, so I can quickly scan the messages, a full page at a time, to see what is important, what can wait and what can be immediately deleted. In fact I can read very short messages in their entirety without even opening them.

I no longer need to carry massive telephone lists from various offices. Through Outlook, I can look up the name, address and phone number of any employee in the company by a number of different methods. I can trace through an organizational chart from the bottom up or the top down.

The implementation of the Intranet also had a profound affect on me personally and in turn all the people we had in the field. Until that time, hardcopy handbooks known as "Sales Guides" were produced and distributed to field personnel for each product or product line. In the case of sales representatives selling our business application products, they might receive one, two, three or even more of these massive tomes, each delivered in 3-inch three-ring binders. The number they needed depended on the number of different products they sold. As Computer Associates acquired more companies with these types of applications, companies like Cullinet, Panasophic, Software International, The ASK Group and others, the number any one sales representative might possibly have in his or her "bag" grew. These sales guides contained product descriptions and status, pricing and competitive information, presentation scripts and any number of other useful pieces of information. When delivered, we also typically included some of those $3^1/_2$-inch disks with presentation material. We tried to update and redistribute them every 6 months, but more typically this was done once a year. In the meantime, as things were changing, as is the constant in the software industry, these sales guides were sitting on shelves growing more and more out of date and obsolete as the weeks and months wore on.

The introduction of the Intranet effectively eliminated the need for these sales guides. This type of information is now posted, not once or twice a year, but whenever a new product or program is introduced, when new tools, information or material is available or whenever something changes. Documents are either viewed on the Intranet or can be easily downloaded as necessary. The trouble we have now is not in making this information available to our people in the field across the globe, but in inundating them with too much information.

One of the ways companies can assist employees in organizing information so they may easily navigate through a potential maze of E-information is through the implementation of a **portal.** There are in fact several different definitions of a portal, some of which refer to particular technology that provides some very specific benefit focused on a particular application, such as a supplier portal you may provide to your vendors. We discuss these types of portals as we delve into E-information shared externally with trading partners. But for our purposes here we are concerned with enterprise information portals. This type of portal is a Web service that delivers many different kinds of information through a single point of access. Therefore, an important feature of a portal is the capability to personalize it to create a tailored workplace so that each user can organize information and applications to work efficiently.

Many different types of information are viewable, including access to an ERP system, to e-mail, to a Web-based news page and other specific desktop tools and applications. The size of the windows can be changed at any time, depending on where the focus of the user has shifted.

This type of portal can be used for much more than a simple means of organizing multiple windows on your desktop. A true Enterprise Information Portal allows multiple employees to work collaboratively with shared information. Let us say multiple employees, in a work group environment, are collaborating on a project where effective communication and sharing of information is critical. Portal users can share this information by publishing and categorizing content in an electronic library that can be accessed by specific subscribers, other work group members. If the Enterprise Information Portal also provides automatic delivery of relevant updated information to subscribers, then the interaction and collaboration between project team members becomes much more automatic. The portal simply needs to have the ability to recognize when an item of information has been added or changed along with the ability to manage the notification and delivery of the updated information.

A very useful added feature of these types of portals is the support of discussion boards or online chats where information is shared in real time with concurrent, interactive dialogues. This is especially crucial today if your workforce is not centralized in one specific location. As telecommuting becomes more common, this feature becomes more necessary.

More specialized portals can be pre-designed for more specific functions, such as sales or service. In these cases the general layout is pre-designed and each user simply customizes his or her view based on preferences. This portal serves more as a menu, with special buttons that have been preprogrammed to take users to applications that are specific to their function within the organization.

Another application of such a specialized portal is in the area of specific employee self-service applications. Reminders and announcements that used to reside exclusively in the domain of employee bulletin boards are now able to evolve to more interactive E-information. In the mid-1990s, progressive companies would notify employees of enrollment deadlines by these electronic means, prompting them to submit necessary paperwork. Today that notification can prompt them to enroll in programs and update their own individual personnel information online. This naturally requires an added layer of context-sensitive security, making employees captive within the process of updating their own information and preventing them from accessing or updating data on any other employee. While this seems like an overly simplistic use of portals and technology, consider today the importance of keeping all pertinent employee data fully up-to-date. September 11, 2001 was a not-so-subtle reminder of the importance of up-to-date and complete contact information. As a company with a New York office in such close proximity to the World Trade Center as to require evacuation of that facility that day, Computer Associates (CA) found out first-hand just how important that information was. As personnel at corporate headquarters on Long Island sought to track down and confirm the location and the safety of each of its Manhattan employees, it became painfully clear how often such information changes and how little thought had previously been given to making sure corporate records were updated. As Sanjay Kumar, the CA CEO, announced in a personal message to all employees that each and every one of the Manhattan employees was safe and accounted for, he encouraged every person to visit the employee self-service portal that very day. The purpose was to ensure that home address, contact and beneficiary information was accurately recorded. While thankfully the latter had proved not to be required for any CA employee that day, had it been and had the information been incorrect or incomplete, there could have been further tragedy added to already tragic losses. Confirming this information without such an electronic self-service method could have taken weeks at a time of fear and uncertainty, not to mention the possibility of additional costs.

There is an old adage, "the shoemakers children have no shoes." Don't let this be your motto by default. As you strive to evolve with E-information, do not focus exclusively on those elements that have direct impact on your public presence. Not only can you derive cost savings by streamlining internal procedures, but you can also derive indirect benefits from protecting and caring for your most important intangible assets, your employees.

COMMUNICATING WITH CUSTOMERS

While caring for your employees has both direct and indirect benefits, many of which go directly to your bottom line, the fastest route to increasing your revenue potential above the line is through effective communication with your customers. Unlike the target audience of your company Web site, the general public, this audience is more select and the information you share is more specific and potentially more confidential. Obviously at this point, security becomes a key issue.

For those companies who market and sell directly to a consumer, there is a significant overlap between the general public and customers. In fact, typically the only difference is that the company has gathered enough specific information about this consumer, such as shipping address and credit card information, to welcome it back to the company Web site with minimal information. But for those conducting business-to-business (B2B), this is an important distinction.

For these companies engaged in B2B, E-information can be enhanced with the introduction of an Extranet. An Extranet acts just like the Internet, but is restricted in access by a user name and password, to a more select audience. This allows more secure access by authorized outsiders to selected information not generally available on the Internet. The most simplistic use of an Extranet is to allow customers to check order status. In the same survey conducted by AMR Research in 1999 that determined that over 90% of companies published a Web site, they also asked the question, "Are your customers or suppliers able to check the status of their order on your Web site?" Figure 4.3 displays the results of this portion of the survey. In this case, we can see that the percentage of those who responded affirmatively dropped to approximately 15%.

Of course, the use of the Extranet can be complemented via e-mail communication to customers as well, possibly notifying customers of order status or notifying them of new information available. When the new information resides in large files, it is far more efficient to e-mail a url, a

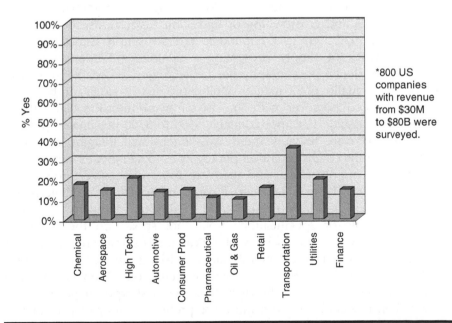

*800 US companies with revenue from $30M to $80B were surveyed.

Figure 4.3   Are customers or suppliers able to check order status on your Web site? Results of survey of 800 U.S. companies with revenue from $30 million to $80 billion. (Courtesy of AMR Research.)

global address of the document on the Extranet, than it is to send the document itself.

The actual E-information you provide to your customers can range from product catalogs, specifications or certificates of compliance to online availability of inventory. In some cases the electronic communication of documents regarding actual transactions such as order acknowledgments and shipment notices might be considered E-information. For those companies not yet actively engaged in E-commerce, that is exactly what they are, but as you begin to engage in the second generation, E-commerce, the distinction begins to blur whether you consider these by-products of E-commerce or E-information.

The effectiveness of a product catalog is determined both by its content and its presentation. If instead of standard catalog items you are presenting configurable products, add a third element to these criteria. The ease of configuring a valid product is also essential. This is actually true whether you offer a traditional print catalog or an electronic version. However, online catalogs can be technology enabled to be more efficient and effective than their printed predecessors. Technology-enabled selling applications not only can provide dynamic graphical details, but also can serve to both validate configurations and suggest complementary products. However, the more actively these applications engage the potential customer with validations and cross-selling or up-selling suggestions, the more we need to categorize them as E-commerce and not simply E-information.

For purposes of simply presenting a catalog as E-information, the technology of digital photography has become very affordable and easily provides high-quality color images. Online product catalogs can be significantly enhanced however with the use of two-dimensional and three-dimensional imaging. For approximately $350 to $500 per image, plus the cost of the software to view them, two-dimensional imaging can allow users to zoom into high-resolution photographs to see greater detail of the product. More sophisticated, and therefore more expensive three-dimensional imaging can be used to rotate, flip and manipulate three-dimensional views to present a full 360° view. Add to this the ability to hyperlink additional details or specifications from "hot spots" on the image or embedded in text and the usability of the catalog is significantly enhanced. It is easy to get a feel for the possibilities by visiting the Web sites of some retailers who have invested in this type of technology, Web sites such as www.godiva.com, www.eddiebauer.com, www.sharperimage.com and www.seibert-rice.com. Or for a more comprehensive tour of possibilities, you can visit www.viewpoint.com and see a demo of a variety of different imaging techniques using two-dimensional and three-dimensional photos.

A fully operable Extranet may, however, involve more than sharing pretty pictures, product catalogs and specifications, particularly where the customers with access are distributors of your product. In this case an Extranet can provide a hub of connectivity between distributors that can have significant impact on the quality of service they can provide to their customers, with the possible effect of contributing to increased revenues for you as well.

CONNECTING DISTRIBUTORS — ANOTHER WAY TO LOOK AT BRICKS AND CLICKS

The Belden Brick Company is a family-owned manufacturer with head-quarters in Canton, Ohio that has been making bricks for over 115 years, and has established itself as a quality leader in architectural brick and pavers, which are available in varying textures, sizes and colors. Belden Brick sells its products to distributors, who in turn sell them to mason contractors or builders. These distributors typically manage building mate-rial yards that would not only carry bricks, but also blocks, sand, mortar and various tools they sell both to the public and also to contractors. Belden is therefore very dependent on the construction industry, fraught as it is with the uncertainty of builders' schedules, the fickleness of architects and the unpredictability of the homeowners' tastes.

Belden Brick implemented an Extranet for the purpose of keeping its distributors informed about their orders, what has been shipped and the brick they have sitting at the plant. Given the typical profile of these distributors, it is easy to see that Belden was not driven to install an Extranet by its customers. In fact, it has some that are very small operations with no desire to use the Internet at all. But about 40% are using it, some very actively. Belden is not pushing its distributors to place orders over the Extranet, it simply views the Extranet as a means of improving communication, and in doing so the ability to operate more efficiently itself.

The type of information that is available over the corporate Extranet is the same information these distributors used to call about. However, Belden does not rely on the distributor to take the initiative to log onto the site to inquire. Belden IT Director Jeff Adams viewed the rollout of this information with the same view toward the way he manages his day. He found he seldom proactively logged into a Web site to find out what was going on. But, like most connected individuals today, he does read his e-mail religiously. Therefore, Jeff supplemented the inquiries available over the Extranet with e-mail notification for order acknowledgments, ship notices, schedule changes and also about specialty brick availability. The distributors then have the option of investigating further by logging onto the Extranet, very much like Dubbed Over's Notify Me List.

Belden's Extranet also provides some additional features, including a picture gallery of the product line and the ability for the distributors to trade computer-aided design drawings. But perhaps the most popular additional benefit to the distributors is the capability, not only to communicate with Belden, but also with each other through a "list server." Belden allows individuals simply to subscribe to the list and subscribers are able to post notices and respond. These notices and responses are then broadcast to all the individuals who access the list server.

If you have ever been even remotely involved in any kind of construction project, whether residential or commercial, large scale or a simple home improvement project, you know that not only do schedules change, but designs and plans are also subject to potentially constant revision. And not only are plans subject to change, but contractors often over- or underestimate the requirements for a job.

The net result can be that the contractor needs an additional number of bricks to complete the project. The distributor has none in stock and therefore must order more from Belden. The customer service representative replies, "No problem, we have another run of that particular brick scheduled in ... 13 weeks." Picture the contractor breaking this news to his customer who just sold her current residence and hoped to be in her new one in 4 weeks. Now chances are another Belden distributor has that brick in stock, since overestimates are also likely. And before the distributor tells the contractor the bad news, and the contractor tells his customer she will have to wait, you can be sure that the phone lines between known distributors will be buzzing.

This is exactly why Belden's list server is one of the most popular features of its Extranet. Instead of making multiple phone calls, and still only reaching a handful of the available distributors, a single notice can be posted and within hours or even minutes, an alternative source can be found. If the quantity required is relatively small, the distributor who can potentially unload the stock from his inventory also benefits, since small quantities that are not sufficient for a typical project can be hard to move. The other distributor is happy to move it out of his yard and off his books. The result is the more efficient satisfaction of one distributor's customer's needs and the reduction of carrying costs and potential inventory write-offs for another.

Belden sponsors no transactions, nor is any E-commerce even conducted. It touches no additional inventory and incurs no additional cost. Belden simply provides a hub through which E-information flows, facilitating improved customer satisfaction and distributor profitability.

CONNECTING WITH SUPPLIERS

Because Internet Procurement has attracted so much interest recently, it is easy to assume that all communication with suppliers is E-commerce.

But this is far from true. If you recall, one of the additional benefits Belden's distributors derived from its Extranet was the ability to trade computer-aided design documents. Now consider participants along the supply chain where more highly engineered products are being produced. Electronically developed design documents can be transmitted either for the simple purpose of sharing E-information or for more sophisticated use in collaborative design efforts. Also consider the transmission of either simple or complicated product specifications or quality standards for measurement.

The transmission of product catalogs from your suppliers does indeed represent the communication of E-information, but usually this maturity level of E-information does not occur until you have begun to engage in some form of E-commerce. Without this level of electronic transactional activity it is just as easy to log onto your supplier's Web site to garner information needed. Viewed from the perspective of Belden's distributors, Belden is their supplier. The fact that their supplier provides an Extranet in order for them to view order and shipment status, and to communicate with other distributors, is a nice benefit of doing business with Belden. Yet nothing more is required from the distributor that wishes to participate in the supplier's Extranet than the ability to receive e-mail and access the Internet. And the distributors are required to send no E-information to their supplier, Belden.

But certainly there is potential benefit to be gained from communicating E-information to your suppliers. Anticipated product demand provides important input to planning that can assist your suppliers in better supplying you. This is E-information that can be pushed to your suppliers even if you are not actively engaged in E-commerce.

An example of one company that has done just that is a company that has asked to remain anonymous. For purposes of clarity I will simply refer to this enterprise as Company XYZ. Company XYZ is a leading manufacturer of commercial and home beverage equipment. It is probably best known for its commercial coffee brewers, but in fact it also has a full line of iced tea and coffee brewers, coffee grinders, specialty drink and hot water dispensers.

However, in this B2B endeavor, the desire of Company XYZ was to link more closely with its suppliers for information sharing and joint planning. Yet Company XYZ recognized that many of its vendors were small companies, which had neither a sophisticated ERP system nor the IT resources to support E-commerce or full E-business integration. The challenge it set was to link more closely with these suppliers while requiring only the minimum of IT resources at the supplier's site.

Company XYZ chose to work with its ERP vendor to design an extension to its planning and purchasing applications. This extension took the form of a secure Web portal, which the company can make available to its suppliers. Through this portal vendors have view-only access to the Company XYZ MRP planning information, as well as to open purchase orders Company XYZ has submitted to them.

Where authorized, the supplier can acknowledge changes to the schedule for open purchase order releases or create purchase order releases against certain periods of the MRP demand. Filters are used both to signal the vendor of high-priority issues and to signal Company XYZ of the vendor's difficulty in meeting the delivery requirements. These signals are delivered in the form of e-mail notifications. The supplier is then able to respond by logging onto a Web application. No specific software, other than an Internet browser, is required at the supplier's location. Yet through this application, suppliers have a view of the data that pertains to them, which resides in the Company XYZ ERP system.

In fact, for its larger, more sophisticated suppliers, Company XYZ provides a means to link directly to the vendor's system, allowing the created purchase order releases to flow back to the supplier's order entry system, thereby enabling E-commerce. However, for many of its suppliers it simply uses the application to enable the exchange of E-information.

THE TRANSITION TO E-COMMERCE

So you might be wondering: When does the transmission of these documents transition from E-information to E-commerce? In the examples we have used so far, where the Belden Brick Company sent e-mail notification of order acknowledgments, shipments, and other documents to its customers, we called it E-information. When Company XYZ pushed new product demand, along with changes to schedules generated by MRP, to its suppliers, we called that E-information. When does the process of sending documents to customers and suppliers stop being E-information and start being important elements of E-commerce? The most significant difference is the level of human intervention associated with these transmitted documents. In the case of both Belden's customers and Company XYZ's suppliers, there was a human at the other end of the transmission, looking at an e-mail or a screen inquiry or printed document, and making decisions. It is when there is a computer on the other end of the line that must read and interpret the contents of the document that we must now transition the discussion from E-information to E-commerce.

E-COMMERCE — THE NEXT GENERATION

Quite simply, E-commerce is the electronic enablement of those transactions, which make up commerce, as we know it today. The final question asked in AMR Research's 1999 survey was, "Are customers or suppliers

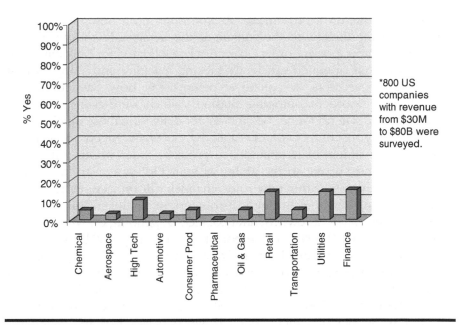

Figure 4.4   Are customers or suppliers able to conduct transactions on your Web site? Results of survey of 800 U.S. companies with revenue from $30 million to $80 billion. (Courtesy of AMR Research.)

able to conduct transactions on your Web-site?" These results are displayed in Figure 4.4. Note another significant drop in percentages. As you transition from E-information to E-commerce there are some important distinctions between operating in the mode of business to consumer (B2C) and B2B. In this context, another important distinction is whether you are considering inbound or outbound communication.

Consider first a B2C environment. You manufacture a product that is sold directly to a consumer. If prior to entering the B2C E-commerce generation you were selling your product through distributors or some other intermediary, or directly to retailers, the implications of selling directly to the consumer can be enormous. For many companies today, this has as many business process ramifications as it does technological. If previously you sold through distributors or retailers, some of which provide their own distribution network, chances are you were dealing with periodic orders based on forecast demand. For seasonal products these forecasts may well have been for an entire season. For products with smoother demand, they may have been quarterly, monthly or weekly. Except for perishable goods, seldom were these orders any more frequent. Because of the periodic nature and the consolidation of individual demand through an intermediary, these orders may well have been for case lots or palettes and your distribution facilities and procedures were probably designed for this level of demand. As you open up a new sales channel for E-commerce directly to the consumer, the demand is no longer gated for you or by you, nor are order quantities consolidated for bulk shipment. Warehouses that operated efficiently picking, packing and shipping bulk quantities of case lots or palette shipments will now have to handle individual piece orders. Shipping methods, which were cost-effective for full truckloads, will no longer be efficient for delivering these much smaller orders. And your order volume will, by necessity, need to increase, potentially by orders of magnitude, to preserve order values. This alone may have some impact on the ability of your ERP system to keep pace.

This presumes that this new sales channel will replace, either in part or in total, your channel through distributors or retailers. This of course raises another issue entirely, channel conflict. Depending on who currently sells your product directly to the consumer, this may not be wise. Your success in engaging directly with your customer will in large part be dependent on the relative dominance you enjoy within the supply chain, as compared with the company that currently sells your product. Nobody would question the dominance of Dell Computer relative to retailers like Circuit City, Staples or Office Depot. However, several of the retail giants today hold enormous power over distributors and manufacturers who are upstream from them in the supply chain. While they cannot prevent you from opening an electronic storefront and competing directly with them, there is also nothing to prevent them from refusing to offer your product in their stores. As a result it is no surprise that the companies that are most likely to venture into the B2C E-commerce game are those companies that are already dealing directly with consumers. Does that mean that if you are currently not engaged directly with the consumer that the intro-duction of E-commerce will not affect you? Not necessarily. Yet that effect may be indirect and may have more impact on the way you fulfill demand than how you actually sell it. We explore these issues further in Chapter 5.

The other opportunity that E-commerce may present to you is the ability to reach new markets that were difficult or impossible to reach through traditional distribution channels. These may be remote geograph-ical areas of your own country that distributors or even retailers found unprofitable or impossible to service. To most urban and suburban Amer-icans this concept may seem as remote as the territories themselves. Yet in other less-developed parts of the world, these areas are very common. Unfortunately in these underdeveloped or emerging economies, the con-sumer is far less likely to have access to the Internet than even in the remote rural areas of the United States. Therefore, the potential for market expansion into these areas, through E-commerce, is relatively limited today.

However, the ability to actually reach global markets virtually overnight is definitely within the reach of E-commerce initiatives. Where it used to takes companies decades to expand regionally, then nationally, and finally internationally, an electronic storefront can reach a global audience almost immediately. For a U.S. company, reaching all English language–speaking countries is literally accomplished instantaneously. I qualify it as "almost immediately" when referring to non-English speakers. The translation of the user interface of most Web-based applications is trivial compared with the implications of conducting business internationally. You must understand and conform to local tax and trade regulations, including customs and import/export rules, in whatever part of the world you conduct business, and you must deal with the logistical issues of products crossing international borders. Just because you are able to market your product globally does not mean you are able to trade globally, regardless of the sophistication of the technology.

B2B VS. B2C

So, what is the most significant distinction between B2C and B2B E-commerce? In conducting B2C transactions, a person uses a Web-based application, an electronic storefront, to enter an order to be processed. This is an interactive process that ends with a sales order being stored directly in that application. The fact that someone besides one of your own employees enters the order produces some immediate cost-savings benefit. However, chances are this is not the order entry module of your ERP system, and therefore there still remains the challenge of getting the order, which now resides in this front-end, outward-facing application, into your back-office system. There are, of course, many different approaches to this, not all of which are automated or even reliable. Most readers today have had the experience of ordering something over the Internet. You type in your name and address correctly, along with the code for the product you ordered. You know that because the online acknowledgment you received displayed it correctly. Then the package arrives and your name is misspelled or the zip code is wrong, or you receive the wrong item. How do you suppose that happens? This is an obvious symptom of using "sneaker net." Someone prints out the orders from the Web-based application and manually enters it into an ERP system. This may cause you to chuckle, but don't underestimate the number of companies that use this approach to "integrate" their outward-facing front-office systems with their back-office ERP application. For the most part it is transparent to the customer, and quite frankly, it works. Are these companies engaging in E-commerce? Sure they are. They just aren't doing it efficiently. You could say, they are still in the toddler stage of this second generation of E-business.

Let us contrast this to what happens in a B2B environment. First of all, the electronic order that comes to you is far less likely to be directly entered by a person using your Web-based electronic storefront. This would require that an employee of your customer's company logged onto your electronic storefront and directly entered the order. Possible? Sure. Likely? That depends on the influence you exert over your customer and the nature of the relationship you have with that company. It is certainly possible where you have a very close and collaborative relationship, very unlikely if you do not. What is much more likely is for the order to arrive as a file, which has been transmitted electronically to you. In many cases, this order will have been generated automatically by a back-office application. This is something most modern ERP systems are able to do in some format. Now that the order has arrived, not only is it not in your ERP system, but it isn't stored in any "system" at all. Would sneaker net work here? I am sure that there is more than one or two companies that have resorted to somehow printing this file out, manually interpreting it and then reentering it into their ERP system. But if so, these companies aren't processing a large volume of these orders and they are not interoperating efficiently with their trading partners. My guess is they won't be doing it for very long.

Sounds a lot like EDI (Electronic Data Interchange) doesn't it? The predecessor, or some might call it the earliest version of E-commerce, was EDI. Back 10 years ago when the big three automakers in the United States first started requiring customers to implement EDI and the larger retailers made it known they would soon follow suit, everyone thought that it would be pervasive throughout many industries long before now. Yet that simply has not happened. Why not?

Constantly changing, multiple sets of standards, coupled with the necessity of expensive private networks, simply made it too cost-prohibitive for most companies. Those who did implement EDI were either very large companies or smaller companies forced to do business this way, or not at all, by large, dominant customers. Many survived this ultimatum by fulfilling only the bare minimum of requirements. The fact that there was implementation effort required at both ends of the transmission lines, for each new company added to the network, was the gating factor in the rollout of this massive effort. Even the automakers and the large retailers were forced to slow their efforts because of the length of time and the effort required to connect new suppliers.

And conversely, just because a supplier was able to meet Ford's requirements didn't mean it could immediately accomplish the same objective with General Motors or Chrysler, since the standards were different. These standards defined the format and structure of the data that was passed along with the communication protocols. It was through these standards that programmers knew which pieces of data to include, which order to place them in, which format and length of the data field was required and what special characters were used to separate pieces of data. All data passed was very structured and subject to these rules. Suppliers to the retail giants faced the same obstacles. So just because a small to medium-sized company had overcome the hurdles once did not necessarily mean it was willing or able to make EDI its dominant or exclusive way of doing business.

For those not familiar with the processes required for EDI, let's examine what it takes for a simple, straightforward order sent from Customer A to Supplier B. The first step in the process is the collection and organization of data by Customer A's internal application systems. Rather than printing out purchase orders, its system builds an electronic file of purchase orders (POs). The next step is to **translate** this electronic file into a standard format. The resulting data file will contain a series of structured transactions related to the purchase orders. Customer A's EDI translation software will produce a separate file for Supplier B. Customer A's computer automatically makes a connection with a **value-added network (VAN)**, and transmits all the files that have been prepared. The VAN will process each file, routing it to the appropriate electronic mailbox for each manufacturer. In this way the VAN serves as a clearinghouse for the electronic transactions — sort of a private electronic mail service. In this way all Customer A's POs can be sent to a single destination, unless of course Supplier B does not subscribe to Customer A's VAN. In this case those files must instead be automatically routed to the appropriate network service, possibly another VAN. Supplier B retrieves the file from its electronic mailbox at its convenience, and reverses the process that Customer A went through, translating the file from the standard PO format into the specific format required by its application software, in order to process the POs.

So how is E-commerce any better than EDI? Enter the Internet and one part of the problem goes away. Expensive private or dedicated communication lines are not needed to participate in these networks; even the smallest companies are able to immediately communicate through the Internet with other businesses. But what about the problem of standards? Without a human reading and interpreting a transmitted document, as is generally the case with E-information, how does this level of automation occur? How does one computer interpret, translate and understand the contents of the document another computer sends to it? We come back to the same answer, standards.

Actually the problem of standards has not entirely gone away. However, a new technology called XML (eXtensible Markup Language) has somewhat simplified the problem. XML is a universal data exchange format. An XML document actually contains data, which is self-describing, a very important characteristic. While EDI relied on an externally documented standard to tell a programmer, or a computer, that the order number was contained in the first ten alphanumeric characters of a file, an XML document "tells" the receiving application where its order number is and how it is formatted. Therefore, there is no need to reformat or massage incoming data specifically for the receiving application. The need is simply to create an XML document.

While XML eliminates many of the problems with standards, there are still some issues that need to be considered relating to the content of the documents to be exchanged. Quite simply put, if your customer sends you an order electronically, and you expect to translate that electronic document into a sales order in your order entry system, that document had better contain everything your order entry system needs to define the order. As a result of these residual issues several new standards bodies have been created.

Several of these relatively new organizations are worth noting. RosettaNet is an open industry group formed to develop cross-industry E-business implementation guidelines around XML. It is actually a consortium of companies from the high-tech industry. RosettaNet distinguishes itself from other standards efforts by incorporating the business process associated with the transaction into the standard. The standard processes, together with the data set required for intercompany communication, are known as Partner Interface Processes (PIPs). Without a standard set of processes supported and implemented within an industry, the cost of implementing standards can easily outweigh the potential benefits. The expectation of this consortium is the emergence of plug-and-play process and partner integration.

To join RosettaNet, companies must agree to the following three requirements:

1. A C-level executive must actively participate.
2. The company must pay a $20,000 membership fee to support research and other activities.
3. The standard created must be implemented.

About 400 companies are currently members of RosettaNet.

So far the organization has published 72 different PIP standards. These document the standards associated with business processes such as the distribution of engineering changes, notification of shipment details, notification of failures, the distribution of product stock-keeping units (SKUs) and the changing of basic product information.

RosettaNet recently joined the Business Internet Consortium to help drive XML standards convergence activities. Within this context both are participating in the E-business Standards Convergence Workgroup, whose charter it is to make recommendations to various existing standards bodies.

As an alternative to RosettaNet, ebXML is a global electronic business standard that is co-sponsored by UN/CEFACT (United Nations Center For Trade Facilitation and Electronic Business) and OASIS (Organization of Structural Information Standards). The ebXML standard is based on published business processes and business scenarios that are common to business transactions. Companies can choose from those published and add scenarios of their own. As they do, the participating companies register themselves by providing a business profile, known as a Collaborative Protocol Profile (CPP). The CPP details the business processes the company supports, the role it plays in the process, the data (messages) it exchanges and the transport mechanism for the messages. From these published CPPs, Collaborative Partner Agreements (CPAs) can be negotiated allowing the companies to agree collaboratively on the exchange of messages that meet the needs of both organizations.

In addition, it may be worth noting that in May 2001, IBM, Microsoft and Ariba jointly launched UDDI (Universal Description Discovery and Integration). UDDI is intended to be a yellow-pages-style B2B directory that would allow companies not only to locate potential suppliers and partners, but also to facilitate linking them up via the Web. There are about 500 companies listed. While the idea is a good one, there is nothing automatic about this as yet, and industry experts seem to agree this will not be a reality for another 3 to 5 years.

OUTBOUND COMMUNICATION — B2B VS. B2C

Before we wrap up this general discussion of E-commerce, recognizing we will delve deeper into the subject in the next chapter, you will recall that earlier I made note that it was not only necessary to make a distinction between B2C and B2B, it was also appropriate to distinguish between inbound and outbound communication, in the same context. I make this distinction because, in general, outbound communication in a B2C environment is far simpler, and could actually be categorized as E-information. Generally an electronic acknowledgment is sent upon receipt of an order. Yet the format and structure of this acknowledgment is entirely left to the discretion of the company fulfilling the order. If sent at all, it is generally delivered through e-mail and as long as the person who reads the e-mail can understand it, it accomplishes its purpose. Shipment notices may also be sent in this manner, but again, humans are available for the interpretation of these documents.

But what about these outbound communications in a B2B environment? It is not unusual for these acknowledgments and shipment notices to be in this same format even in a B2B environment? Belden Brick was issuing these types of notifications via e-mail. Yet if you recall, Belden Brick was not engaging in E-commerce. As your implementation of E-commerce matures, less and less human intervention is required, meaning your E-commerce applications must be able to communicate and interoperate with the E-commerce applications of your trading partners. In today's environment, this usually can be interpreted to mean that in issuing these types of acknowledgment, shipping and invoicing documents, that you will be generating them in XML format.

AUTOMATING PROCESSES

The ultimate level of maturity you reach in terms of E-commerce generation will be determined, not only by the transmission of documents, but also by how much of the full flow of business processes you have both automated and electronically enabled. The company who receives electronic orders only to manually enter them into another system has not evolved very far. Additionally this is only one step in the full order-to-cash cycle. Notification of order acceptance and shipment are logical next steps, but the path of evolution continues beyond to invoicing and receipt of payment. All these steps in the business flow can be electronically enabled and automated to the extent that little or no manual intervention is required. Typically these functions are performed by your ERP system, and many ERP systems today have been enhanced to electronically enable them. Yet if the electronic enablement can only be performed using EDI, your next logical step might be the enhancement of these functions to enable them to produce XML documents. Your choices may include either directly enhancing your back-office applications or possibly purchasing and implementing specific collaborative commerce applications that serve as a bridge to full and flexible interoperability.

The extent to which you are able to support collaborative efforts, automate and interoperate determines the maturity of E-commerce in your company.

FULL E-BUSINESS INTEGRATION — THE FINAL FRONTIER

As you mature through the first two generations of E-business you find yourself working within your existing business model. E-information simply supports the business processes that have brought your company to its current level of success. Neither Belden Brick nor Company XYZ was fundamentally conducting business differently as a result of its efforts to better communicate information electronically. Your basic business model need not change when you progress to conducting E-commerce, although we saw the potential for this if in fact you desire to engage directly with consumers, and have not done so in the past. However, the typical effect of E-commerce is not to change your business processes, but to streamline and automate them.

This changes quite dramatically as you approach the third and final generation, full E-business integration. At this point you begin evaluating and considering changing the fundamental way you do business. In Chapter 1 we introduced the concept of virtual enterprises and virtual integration. Companies today have become less vertically integrated in attempts to focus on core competencies. This has necessitated new ways of doing business, as we depicted back in Table 1.2, and has some very specific ramifications in our economy today. The value chain has lengthened and become more complicated; yet expectations of response time and delivery performance have risen dramatically. New business relationships are formed in order to reestablish the ability to monitor and control the entire process, while spreading and reducing the risk factor.

Therefore, as you approach this third and final level, the potential increases dramatically for your underlying business model to change.

HOW DOES E-BUSINESS CHANGE YOUR BUSINESS?

There are as many different answers to the question, "How does E-business change your business?" as there are ways for your business to be unique. Gary Layton, Vice President of Marketing for Computer Associates, likes to use a snowflake analogy to describe BizWorks, a software product that enables E-transformation. According to Gary, just as no two snowflakes are the same, each implementation of BizWorks is different, solving unique problems and providing return on investment individually to each company.

Gary's analogy is particularly useful, not only because it succinctly describes the product he markets, but also because most companies like to think of themselves as different from the rest. I personally like it because of what he doesn't say about snowflakes. While hypothetically no two snowflakes are alike (I say hypothetically because I have often wondered how you could ever conclusively prove that true), you only have to step back a short distance and they all look alike. After all, they are all made of the same stuff, frozen water. Similarly each business is made up of common processes. This is true both on the manufacturing floor as well as in the back office. If you make food products, you mill, grind, chop, blend, stir, cook and/or package. If you make plastic components, you heat, extrude, mold, trim and/or assemble. If you make electronics, you wire, solder, assemble and/or test. In all these industries you receive materials, inspect, put materials away, then pick, pack and ship parts. In your back office you process sales orders and generate purchase orders. You match receipts to invoices and pay your vendors. You invoice your customers, receive cash and apply payments. You balance your books and produce income statements and balance sheets.

I may be oversimplifying your business, but in many ways all businesses are made up of the same stuff, just like all snowflakes are made from the same stuff. If they weren't, no ERP system would ever have been sold to hundreds or thousands of companies.

I referred to myself before as a manufacturing consultant. I used the term loosely. For the 10 years I may have called myself this, in fact I was seldom performing consulting services as a management consultant. I was presenting and recommending software that modeled and supported business processes. I found that by knowing what the company manufactured and by spending less than an hour touring the factory, stockrooms or warehouses, and the back offices, I could flowchart the company's business processes and be guaranteed at least a 75% level of accuracy. One or two additional days conducting interviews with key personnel could bring that up to 85 to 95% accuracy. Like snowflakes, without a microscope, they all appear similar. After all, snow is snow. It's all cold, wet and white. But drawing the conclusion from this that all companies are alike is like saying every snowstorm in New England is the same. Just a 20% difference in the combination of temperatures, water content and wind speed can make the difference between peaceful, light, beautiful snow and deadly and destructive slush and ice. And the same is true for businesses.

All companies comprise similar components. That is, they conduct similar business processes. Yet how the processes are performed and the myriad ways in which they can be combined is what makes each business unique. As a business is transformed into an E-business, some of these processes, or the way they are combined, must change. The business processes fundamental to any business must be performed, but the who, what and how can change very dramatically. Therefore, to presume that I can tell you exactly what will change and how is unrealistic. About all I can guarantee is that through E-transformation, the velocity of your business, along with the pace at which you will be confronted with change, will both increase. And to better prepare you for these changes I can provide you with some examples.

THE STORY OF DELL

Most of the companies I use as examples in this book were included because I have had some level of direct personal experience with them. However, in choosing a company as an example of how businesses can change in the context of E-business, no company provides more insight into this or epitomizes the opportunities better than Dell. Yet my personal experience with Dell is limited to the use of a Dell laptop and the opportunity of hearing Michael Dell participate in a panel discussion on leading multiple technology initiatives at a recent CEO conference.[1] However, it is not difficult to learn about Dell, since it has achieved a degree of full E-business integration few others have. As a result, Dell is widely referenced by industry analysts and journalists alike. In addition, much of what I learned was made possible using the 32-page catalog I received, unsolicited, from Dell in the mail.

Michael Dell transformed not only his own company, but an entire market. Prior to 1993 nobody, except for Michael Dell, believed there was much of a market in selling PCs directly from the manufacturer to the consumer. Most believed that, like the mail-order clothing business, it would never amount to more than 15% of the market. While the first 6 months of 1993, using this direct-sales approach, resulted in a $65 million loss, 6 years later direct buyers accounted for nearly a third of the PC business, to the benefit of not only Dell, but Gateway and Micron Electronics as well. At this point, Dell is the clear and undisputed market leader in direct sales of PCs. It has turned a low-margin mail-order operation into a high-profit, high-service market-leading business.[2]

The reason for this is clear. There is much more going on here than just E-commerce. The not-so-well-kept secrets to Dell's success lie in speed, tight integration with suppliers, keeping inventory low and in shaping demand instead of relying on forecasts. Add in a strong dose of globalization and you have a winning formula.

Dell has long been a proponent of just-in-time inventory within its own walls, but it further accelerates the supply chain by also applying the same requirements to its suppliers. Dell reduced its number of component suppliers from 204 to 47 and collaborates with those to make sure components are warehoused just minutes away from Dell's factories in Austin, Texas, Limerick, Ireland, and Penang, Malaysia. In some cases it has outsourced selected business processes. In the United States, Caliber Logistics manages a warehouse 15 minutes from the Austin assembly site. Suppliers restock the warehouse and manage their own inventories. Caliber supplies Dell directly as parts are needed. Dell is billed for the components only when they leave the warehouse.

And then there are monitors. Dell no longer receives deliveries for these components, but instead sends e-mail to a shipper such as United Parcel Service when a machine is ready to ship. The shipper pulls a monitor from supplier stocks, much as Caliber does, and schedules it to arrive with the PC. This is an example of combining point-of-consumption inventory principles, employed primarily by larger retailers, with third-party logistics and vendor-managed inventories.[2]

You may not have the dominance within your supply chain to demand this kind of collaborative effort from your suppliers, but you may very well some day find yourself in the same position as Dell's suppliers. Those who are not able to respond may find themselves in the same position as the 157 vendors who were eliminated from Dell's supply chain. Those who survived were able to collaboratively integrate their business processes with those of Dell's. This requires the ability to respond not only in terms of velocity, but also in terms of integrating back-office systems. We discuss integration of disparate business systems in much more depth in Chapter 6.

The advantages to this approach are clear. Turning inventory fast has always been an elusive goal of many manufacturers, as well as retailers. Inventory serves to impede product innovation, and in a market such as Dell operates in, this can signal death as fast as a speeding bullet. Keeping inventory low is the most reliable means of responding to market demand and customer preferences.[3]

Dell's business model is not only one that is consumer-direct, but also one that is build-to-order. Most manufacturers of consumer products use a much more make-to-stock approach to the manufacturing process and a make-sell-ship approach to their business. Dell is a pioneer among those companies that have sought to shake up that most basic of business models. They have turned the tables and use instead a sell–make–ship model. Forecasting customer demand has long been the secret to success in balancing inventory with customer service. This is true especially in a make-to-stock business. Not being able to predict consumer demand accurately is the source of price markdowns, liquidation sales and inventory write-offs. Yet the transition to a make-to-order model does not preclude the need for accurate forecasts unless, like Dell, you are able to procure all the necessary components within hours, or perhaps minutes of seeing actual demand. But even companies, like Dell, that enjoy this type of collaborative relationship with their suppliers feel the need for accurate forecast of demand. Otherwise, how can their suppliers provide them with adequate service levels and still stay in business? Pushing the risk of inventory upstream in the supply chain only works if the suppliers themselves can profitably strike a balance between inventory and customer service. And you can imagine Dell's expectations for customer service levels.

According to Randy Mott, CIO of Dell, his company has not excelled in forecasting customer demand.[4] And yet, you can order a custom-configured PC at 9 A.M. Monday and it can be on a delivery truck at 9 P.M. on Tuesday. If Dell has not mastered forecasting demand, how does it do that? Instead of forecasting demand, Dell seeks to shape that demand. If you browse through the Dell catalog, or you seek to order your computer online, you are presented with a variety of sample configurations, which are very easy to personalize. For telephone sales, Dell can alter its suggested configurations daily in order to guide customers to PCs with generally available parts. And as many as two thirds of all telephone buyers accept recommended configurations. And for larger corporate buyers, which account for 60% of its sales, it holds annual technology briefings, picks the customers' brains on needs and issues, and then designs standard configurations. CA is one of these corporate customers. An employee of CA has only to bring up an online form and identify his or her job function, and the employee is presented with a standard configuration of a desktop or laptop.

So by keeping inventory low, building responsiveness into its supply chain and managing consumer demand, Dell has achieved an almost lethal advantage in innovation and in pricing by capturing premium prices early in every product life cycle.[4]

But Dell doesn't stop at its component suppliers in forming the kind of virtual enterprise we described in Chapter 1. Just flip through its catalog and you see other types of partnerships. Not only can I configure an Epson or HP printer with my custom configuration, but I can enhance my system by upgrading from Microsoft Works to Office XP. I can also have my PC come preconfigured with Dell Jukebox powered by MUSIC-MATCH®, Internet Explorer and Norton Antivirus software. I can purchase 6 months of Internet Access either from America Online or through DellNet™ by MSN™. And I can pay with credit card or direct payment, or finance my purchase with a 48 monthly payment plan offered by CIT Online Bank. But I do all of this through Dell either over the phone or online, as if everything I was purchasing was "manufactured" by Dell. In this way I have a virtually integrated buying experience and rely on Dell to put all the pieces together seamlessly.

PREPARING FOR FULL E-BUSINESS INTEGRATION

Dell serves only as an example of full E-business integration, because the definition as it applies to your business must be established in the context of just that — your business. There are, however, some key characteristics of all true E-businesses. First of all, the speed at which they operate is faster than the speed of traditional business. As a result of this increase in velocity, there is an increase in the pace of change. And, finally, the importance of interoperability within an integrated business community is paramount. Knowing this, there are certain steps you can take in preparation for this third and final stage of evolution.

The first step you can take is to improve your business flexibility and your ability to adapt to changing business climates as well as changing business processes. As usual, technology in general, and information technology in particular, plays a key role in this preparatory step. Chances are the software that you use to run your business today will not be entirely sufficient to transform your business. That means you will be shopping for new software, which either complements or replaces software that you own today.

As you evaluate different business applications and tools from software vendors, listen carefully to what these software vendors say. Do they talk more about working with other applications, or do they talk about replacing them? Whether or not your existing software needs to be replaced or supplemented is another whole topic of conversation, which we address more specifically in Chapter 6. But even if you could replace everything you own and start from scratch, the applications you run still need the ability to interoperate, at some level, with applications owned by trading partners. Collaborative supplier relationships such as Dell has established with its component vendors are not easily supported with phone calls, faxes and manual intervention. In order for Dell to ship your order for a custom-built computer 36 hours after you order it, the integration of all the component business processes must be seamless. This is the hallmark of full business integration.

Software vendors of choice should be those with strong interoperability support. These are vendors that can connect to heterogeneous systems and networks, and are able to exchange data in a wide range of formats. The importance of this cannot be overstated, as integration is costly and adds little to the value of the applications themselves.

AN ANALYST'S PERSPECTIVE

Bruce Richardson, Senior Vice President of AMR Research, suggests, "2002 will mark the 'Era of the Real-Time Enterprise,' meaning that companies will explore how to open their systems and business processes to customers, suppliers, and other trading partners to make them part of the extended enterprise. This is the central tenet of our work in Enterprise Commerce Management [ECM]."

AMR Research has developed the concept of ECM as a blueprint for companies to make sense of the broad range of business applications they may have acquired, how best to deploy them and how to maximize benefits received. ECM is not an application or a software category but is meant to provide a technical standard for software vendors to meet.

ECM is most relevant in the context of a technological discussion, since in many ways this definition blueprint is a guide to the technology requirements to support my definition of full E-business integration. Although I have made every attempt to translate this definition into lay

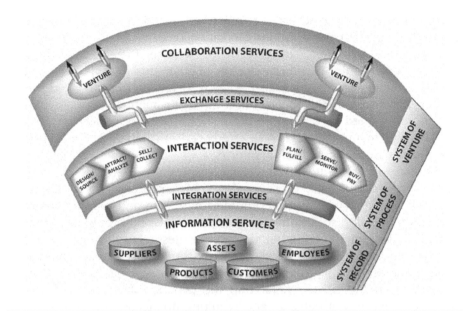

Figure 4.5   The AMR Research Enterprise Commerce Management (ECM) model.

terms, those readers who have little interest in delving more deeply into this level of technological discussion should feel free to skip directly to this chapter's summary.

Figure 4.5 is a graphic depiction of ECM. It consists of five layers made up of three critical sets of services and the technology that links them together.[5]

Information Services

The information services layer forms your corporate system of record. AMR Research is quick to point out that multivendor environments are a reality. There is evidence of hundreds of ERP systems across multiple business units, E-commerce, supply chain and customer applications and homegrown and legacy systems all delivering functionality. For every legacy system that large enterprise applications replace, 30 survive the implementation. The ECM blueprint does not necessarily seek to replace specific applications, but it does seek to provide a *logical* unified view of information, business rules and results, even though the data itself may be physically stored in different databases and even on different computers. This is conceptually very similar to the goal of Steve Myers, the CEO of Myers Industries, as described in Chapter 2. The challenge is essentially one of integration.

According to AMR Research, this layer would provide the services needed to ensure consistent representation of data about products, customers, assets, suppliers, and employees. Whether there is a single centralized copy of any piece of data or whether this logical view simply provides the illusion of centralization is of far less concern to the decision maker than is the ability to present it in such a manner. Myers achieved this logical unified view through the use of a common object model and the wrapper technology that was referenced in Chapter 2. Various other alternatives are discussed in Chapter 6.

In spite of various technical alternatives to unifying the view of potentially decentralized data, there is also a high probability that disparate business applications will be required to pass data and transactions back and forth. In doing so, transactional integrity must be preserved. In order for this to be guaranteed, a certain level of messaging must be available, that is, the ability for one application to communicate with another in such a manner that it "knows" that information sent has been received. This conceptually is very similar to the process of sending an e-mail message, a letter or a package and requiring notification of its receipt. Of course, whenever transactions are passed from one application to another, there is the possibility that the pass will be incomplete. Therefore, interface mechanisms that ensure transactional integrity must have the ability to reverse completed or partially completed transactions.

Another important information service required is event listening and triggering, that is, the ability of the software to detect changes in data elements so that it can trigger events. The software may be "listening" for an event such as the creation of a new order, or it may simply be comparing and correlating data elements. The variety of events that can be triggered is endless. Typical triggered events might be the addition of a task to someone's "to-do" list, or perhaps an alert sent to a desktop or a wireless device.

Interaction Services

According to AMR Research, interaction services form a company's system of process. The research has seen the pendulum shift from decentralization to centralization, back and forth many times. To make companies operate in a more entrepreneurial way, decision making is shifted to the business units. After some period of time, there is concern that opportunities for economies of scale are being overlooked, and there will be a shift back to centralization. And so it goes, back and forth.

The emergence of Global Process Management (GPM) applications provides the possibility of providing a consistent set of rules on an aggregated basis while still allowing some measure of flexibility. GPM applications include the following:

- Supply Chain Management
- Product Lifecycle Management
- Customer Relationship Management
- Workforce Relationship Management
- Asset and Financial Management

AMR Research writes, "GPM applications should have a component design including a meta information model and the ability to expose processes for use in specific instances."[5] While this statement is significant, I am afraid its full meaning might be lost on the nontechnical reader. Therefore I will attempt to translate. Traditionally, systems used to manage business processes, such as those listed above, have embedded business rules within the logic of the application. In this case, the applications impose limits on the agility of the business to change decision-making policies. Removing the rules from the applications themselves and managing them in a separate rules engine paves the way for greater flexibility and creates a more agile environment. When the rules change, the policy criteria, such as thresholds and limits, can be changed without having to touch the actual application logic. In this way rules can be shared across the corporation or be defined for each business unit, depending on which direction the pendulum swings. New rules can be added without replacing or modifying existing applications, and applications can be removed and replaced without disruption to the definition of the rules. This is an extremely important aspect to the implementation of full E-business integration, and represents a significant departure from the traditional design of business applications.

Integration Services

AMR Research considers integration services the connective tissue between information and interaction services. Integration must focus on preserving the integrity of the systems and it must be bidirectional. Although the Internet has significantly lowered the economic barriers to interoperability, connecting outside your own business at both the content (data) and business process level remains prohibitive for many companies. AMR Research feels this is primarily due to software vendors that view their applications as the center of the user's universe, and assume all other applications revolve around theirs. The ECM blueprint puts the enterprise at the center of the universe and demands that interoperability be the responsibility of the software vendors. This means software manufacturers should build products as components of a corporate ECM system, not as stand-alone applications. This encourages a plug-and-play mentality that to date has not been pervasive.

Collaboration Services

AMR Research describes this layer as the means of delivering what it calls a system of venture. It allows companies to embark on strategic, collaborative business ventures with external constituents. Put in the context of our definition of full E-business integration, it allows companies to work cooperatively and collaboratively with other businesses in a manner that is seamless and transparent to the consumer.

The fictitious example frequently used by AMR Research is a Father's Day promotion, a joint venture between Turtle Wax, a manufacturer of car care products, and Home Depot, one of the largest home improvement retailers in the world. Several Turtle Wax products are combined into a single package, including a greeting card that promises a car wash. The two primary partners, Home Depot and Turtle Wax, will collaborate on the package design. They will outsource fulfillment to UPS Logistics, create a cash flow based on purchases and have UPS provide further services to break down packaging and return unsold product. Figure 4.6 depicts the various players in the joint venture.

While on the surface this seems to be a fairly simple promotion, in fact today it would take most companies months of planning, and even

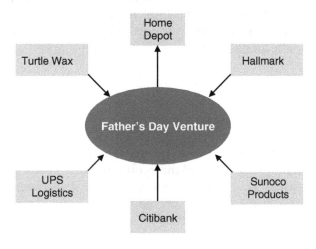

Figure 4.6   Father's Day promotion players. (Source: AMR Research.)

with careful planning there would be little integration of systems among the various players. Yet ventures like these are relatively short term. Father's Day certainly does not even have the "season" length that other holidays like Christmas possess. Therefore, the vision of ECM is not only to integrate the processes, but also to automate them in a very short period of time. For this to occur, selected applications in place at each of the constituents of the venture must have access to processes exposed as Web services from both the primary participants as well as other third parties. Ideally, the venture would also include some integrated analysis, in order to determine its relative success and the potential for reuse.

Obviously there is no set of applications that is designed and delivered specifically for ventures such as this. Business managers must assemble the processes needed to operate the ventures and technology must be put in place to enable them in a cooperative and integrated way.

Collaboration within a virtual enterprise is not, however, limited only to *ad hoc* ventures such as the Father's Day promotion example. It may simply be the natural result of outsourcing various business processes in an effort to concentrate on core competencies and achieve a level of efficiency not necessarily accomplished by traditional vertical integration. The Appendix provides a summary of a survey also conducted by AMR Research on the expected and achieved results of outsourcing both financial and administrative, as well as operational, processes.

Exchange Services

AMR Research portrays the exchange services layer as the bridge between the interaction and collaborative services layers. Exposing internal business processes to external parties for the purpose of cooperative ventures requires a platform for collaboration that also secures internal resources. AMR contends that a Private Trading Exchange (PTX) is essential to these objectives. (See Chapter 5 for a full discussion of trading exchanges.) The PTX then becomes the operating environment to facilitate this interoperability. AMR proceeds to outline the following key services as necessary elements of the PTX:

- **Identity Management** — This includes the ability for both internal and external parties to be identified, along with the roles they play and what they are permitted to see and do over the exchange.
- **Content Management** — Content may range from relatively static data such as product catalogs to dynamic data such as pricing and availability of interactive data such as orders, acknowledgments, shipping notices and invoices. The content may be structured data, such as all these examples, but may also extend to unstructured

content such as multimedia, drawings and those business rules discussed as interaction services.

■ **Integration** — There must be some means of connecting the PTX to the back-office systems of the external parties. As full E-business integration matures, integration becomes tighter and more timely. The end result is seamless and real-time integration.

■ **Process Management** — Pricing, negotiation, returns and promotion management are just some of the examples of collaborative processes that must be managed. Each of these processes is modeled with business applications, whether they are back-office systems or outward-facing collaborative commerce applications. The ability to manage the flow of information and activity between these different applications is the basis of this process management.

■ **Analytics** — Vast amounts of data can potentially pass through the PTX. The ability to capture and analyze this data is necessary to provide performance measurement and can also be particularly useful in identifying trends and predicting the future.

AMR Research's clients can be split into two categories: the end users of technology, seeking advice in procuring and implementing it; and software vendors that provide the technology. AMR has received a generally enthusiastic response from its user community regarding its ECM blueprint. Now it is in the process of working with its vendor community to deliver the necessary functionality to facilitate its adoption. The reader should understand that at this time many software vendors have a great deal of work to do in order to become ECM compliant. However, there are selected vendors with advanced technology that is allowing pioneering companies to achieve the kind of full E-business integration that is still in the future for the vast majority of enterprises.

SUMMARY

As we have seen, E-business represents a whole new orientation of thought. While the term is pervasive today, the adoption of E-business is sometimes far overestimated. The journey to full E-business integration is a long and complicated one, and there are many milestones to be achieved along the way.

The three generations of E-business — E-information, E-commerce and full E-business integration — are truly like the generations of a family. While the birth of a new generation requires certain achievements to proceed from one stage to another, each generation continues to mature even after passing to the next. It is necessary for every business therefore to periodically assess where it stands in terms of acquiring and implementing the technology necessary to interoperate cooperatively and collaboratively with other companies as a member of an integrated business community. The hope is that this chapter has given you some insight in where you stand in the evolutionary process and the path of your journey to full E-business integration.

REFERENCES

1. CEO2CEO Conference Strategic Leadership in Extraordinary Times, October 30–31, 2001, New York, sponsored by *Chief Executive Magazine.*
2. McWilliams, G., Whirlwind on the Web, *Business Week*, April 7, 1997, pp. 132, 134, 136. Updated on July 30, 2000 by R.E. Peschke (pesche@mnstate.edu).
3. Martin, R., Responsiveness to Customer Demand was the Overriding Message at the AMR Research CPG-Retail Conference, available at www.amrresearch.com/research/Alerts/010322sapstory1.asp.
4. Girard, G. and Covill, R., AMR Outlook: Leverage Your Supply Chain to Win the Customer at Every Shelf, available at www.amrresearch.com/research/Alerts/010322rasstory1.asp.
5. Parker, B., Enterprise Commerce Management: The Blueprint for the Next Generation of Enterprise, available at www.amrresearch.com/research/Reports/EMS.0106emsbody1.asp.

5

E-COMMERCE — ONE STEP ALONG THE JOURNEY TO FULL E-BUSINESS INTEGRATION

Chapter 4 described E-commerce as the second generation of E-business and the electronic enabling of the transactions that comprise commerce between trading parties, whether these are businesses or consumers. Although E-commerce is just one step in the evolutionary journey, the strategy you employ at this level can have a major impact on your ability to maximize profits in striving for the ultimate goal of full E-business integration. This chapter discusses the different business models you might employ and the interdependencies among these models, your business processes and your information technology.

CHOOSING A BUSINESS MODEL

Several years ago, as the dot-com frenzy whipped to a crescendo, pioneering companies adopted new business models centered on conducting business exclusively over the Internet. Now, as the business world transitions back toward more sane market valuations, an appreciation of more traditional buying and selling values and the recognition of challenges in deployment of E-commerce initiatives, we see companies approaching E-transformation more cautiously. Yet even the most cautious companies see the writing on the wall or, rather, E-commerce on the horizon. There are efficiencies to be gained from the automation and electronic enablement of these trading functions, and as more and more companies embark on the journey, the less likely it is that you will be able to resist. As more enterprises evolve toward conducting business electronically, they will be less inclined to do business with those companies who cannot. Those "bricks and mortar" companies who lag behind in evolving to "bricks and clicks" entities will find their markets shrinking. There will simply be fewer companies willing to conduct business in this fashion.

Exactly how you go about conducting E-commerce is not as straight-forward, for this decision is based on the business model you intend to use in implementing your core E-commerce strategy. The complexity arises because this is not a singular decision. Your business model for procurement may be very different from your sell-side business model. Even within the domain of procurement, how you approach indirect purchase of MRO (Maintenance, Repair and Operational) materials may be significantly different from the procurement of direct materials, where product complexity may be more variable and markets are more vertical. On the sell side, the product complexity and market fragmentation you face in your sales environment may be entirely different, and may not be consistent across all your product lines.

The preemptive question you must ask yourself before beginning to execute a sell-side E-commerce strategy is whether to use a direct-connect approach or a business model that uses intermediaries. Several years ago, there was much discussion of disintermediation, the elimination of the middleman. After all, those distributors, which served as intermediaries between manufacturers and retailers, attached additional cost, but added little value. The extent to which disintermediation has actually occurred is much smaller than predicted. While distribution networks may add little value to the product that ultimately winds up in the hands of the consumer, they do serve a very important role in the supply chain. They absorb risk. As retail giants, who carry enormous weight in the industry, seek to carry less and less inventory, yet demand quicker response to orders, they essentially push these carrying costs and the associated risks upstream. Without the cushion of the distributor, these risks are pushed directly back to the manufacturer, many of which are too small or too fragile to absorb it.

Yet the choice between a direct link and intermediation has taken on a new dimension in the context of E-commerce, for in this context a new and different type of middleman can be introduced, the E-marketplace or independent trading exchange. In other words, do you offer your products through your own Web site and electronic storefront, or do you offer them through an electronic marketplace? Do you make a direct B2B link with your suppliers or do you seek to leverage the purchasing power of an exchange to aggregate your purchases within a business community? And to further complicate matters, we also see the emergence of a different type of exchange, that of the private trading exchange.

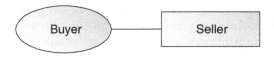

Figure 5.1 Direct sales.

DIRECT SELLING

Direct selling over the Internet is the most simplistic E-commerce business model, as can be seen in Figure 5.1. However, it can be the most complicated to administer and maintain. Unlike an inter mediated exchange, the direct seller must supply the domain expertise, manage the presentation, automate the process, and administer all the components of operating the Web site. In addition, the buyer must be drawn directly to the specific electronic "storefront," and as discussed in Chapter 4, there are no "Fields of Dreams" in the world of E-commerce. Drawing visitors to your electronic storefront is as much of a marketing challenge in cyberspace as it is in the "real" world.

This may or may not be an issue for you, and is primarily dependent on whether you view the Internet as an additional sales channel, for the purpose of extending your market reach, or whether you view it as an alternative method of interacting with your existing customers.

This model is appropriate where there is a low level of product complexity and the seller has an established market presence. Manufacturers will directly interface with another business, or directly with the consumer. This model is far less effective when there is a high level of market fragmentation in an industry with a large number of participants each striving to gain awareness, along with mind share and market share.

However, if appropriate for your business, let us examine the requirements you must satisfy to facilitate this business model:

- You will need to provide order management for both incoming and outgoing orders, which is Web based. It is important to note the difference between "Web enabled" and "Web based." The quickest way for any application to be adapted to the Internet is to make it accessible by a browser. This has many advantages, both Intranet and Internet based. However, it requires that users of the system, whether they are internal or external, have direct access to the application and database servers on which the client's production system is running. While security is controlled through the applications, this only works as long as the user stays in the application. But once a hacker has access to your server, there is a lot of damage that can be done. As a result, security issues remain as long as the application is simply Web enabled. A truly Web-based system allows you to keep your customers safely outside your firewall.

■ It must be truly intuitive — not just intuitive navigation, but the application itself must be intuitive because you will not necessarily have the luxury of being able to train the users of the software.

■ It must have a different level of security than is associated with internal order management systems. It is no longer sufficient to limit the use by a simple password procedure that prevents unauthorized users from running the program — there must be context-sensitive security that limits users to their own orders, perhaps specific products or inventory. And, as noted above, it must be manageable outside a company's firewall.

■ It must accommodate the management of existing standards and emerging standards, such as XML.

■ It must be able to plug into existing back-office transaction-based systems — your own and possibly that of your customers.

■ The second generation of direct selling over the Internet will allow for personalization. This personalization involves applying knowledge about existing customers and products in order to present the appropriate product for the customer's needs, along with the ability to cross-sell and up-sell. This is best accomplished by applying predictive technologies to the existing data available from your information-rich ERP system. We discuss predictive technologies in more depth in Chapter 7.

In a direct selling environment it is important not only to consider the acceptance of an order, but also the support structure you provide to your customer as the order is entered, administered and managed through the delivery process.

When selling directly to a consumer, the relationship you have with your customer can begin and end on the Web. However, in a B2B environment, both initial and continued sales are built on relationships established with customers. Sometimes these relationships are decades old and have been cemented in place using paper, phone or face-to-face meetings. Yet regardless of how strong these have been in the past, they can become fragile at any point without the proper diligence paid to maintaining these relationships. When implementing B2C sell-side E-commerce initiatives the objective is often cost driven. Providing service and support over the phone can cost companies anywhere from an average of $5 to a whopping $80 per call. Replacing much of this through Web-enabled self-service can reduce these costs to 20 to 80 cents per call. Yet the objective of self-service in a B2B environment cannot be entirely cost driven. Instead, it needs to offer a new and additional support channel or some level of customer convenience and hence customer satisfaction are largely at risk.

It is therefore critical that these E-commerce sell-side initiatives, and the resulting support ramifications, be viewed not from your own organization's point of view, but from your customers' perspective. As a result, both integration and knowledge management play key roles. It is not unusual for the costs associated with these two aspects to far exceed the cost of whatever business application you use to accept the order.

Consider the different systems an account manager or a customer service representative might have to access to respond to order or account status inquiries. He or she may access an order management system and an accounts receivable system, both of which may be elements of your ERP system, or stand-alone or interfaced modules. Yet are contracts and service level agreements (SLAs) administered in the same systems? What about service and support calls? It may be that your employees pop in and out of multiple applications during the course of a call or an investigation, but you cannot expect your customers to do the same.

And then consider the knowledge your sales or support representatives bring to the party and consider if that knowledge can be replicated in a self-service environment. Does that mean prompting and guiding your customer through a variety of predetermined paths? Does it mean supporting natural language processing of questions they may pose or providing a series of structured key word searches through a database of frequently asked questions or anticipated problems? The less you are able to accommodate these alternatives, the more likely you will need to be able to connect the inquirer back to a real person, available through an online chat, or via an immediate phone call.

These are simply the requirements to compete, not necessarily to excel in this arena. And if these requirements appear daunting, don't forget that this is but a single business model.

THE INTERMEDIATED BUSINESS MODEL

What is rapidly becoming a more common E-business model is that of the intermediated online trading exchange. While many industry observers of the Internet predicted that it would bring about the demise of the "middleman," allowing businesses and consumers to "buy direct," this has not proved true. However, the value added by the intermediary has and will continue to change. Early marketplaces relied on market aggregation as their primary value proposition and simply matched buyers to sellers

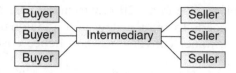

Figure 5.2   The intermediated business model.

in many-to-many relationships as depicted in Figure 5.2, but left many details untouched.

There are various ways of defining or describing online marketplaces. First of all they can be categorized as either horizontal or vertical exchanges. Horizontal exchanges offer products to companies across multiple industries. Vertical exchanges focus more specifically and typically serve a single industry.

Those that are horizontal are broad-based exchanges. They facilitate specific business processes, such as online procurement, but they serve any number of different industries. These, quite simply, are the online equivalent of the shopping mall. The organizations operating these exchanges assist in managing the relationships between suppliers and customers, and manage the E-marketplace. But if they offer domain expertise it is specific to the business process, rather than specific to the products offered or an industry served.

The value such a horizontal trading exchange brings to bear is directly proportional to the size of the community of participants. Like the shopping mall that offers the most convenience to shoppers who can get the largest portion of their shopping done without having to travel to another store location, buyers are drawn to the exchange by providing them access to a large number of the potential suppliers who can satisfy their needs. Stores are drawn to the malls where there is the most traffic of shoppers. Sellers are drawn to the exchange based on the number of buyers who are attracted to the site. Therefore, the most successful horizontal exchanges will be those that can capture the most momentum.

Two excellent examples of these exchanges are the Ariba Supplier Network, operated by Ariba, Inc., and MarketSite.net, operated by Commerce One, Inc. Originally concentrating in MRO or indirect materials, they have been successful in drawing a wide selection of buyers and sellers. Both of these companies saw an opportunity early on to develop what might be called an E-commerce platform or infrastructure — a way to match buyers and sellers, while facilitating actual E-commerce transactions and providing functionality specialized to facilitate the business process of procuring indirect materials. Yet both of these pioneering companies had to build these E-commerce platforms from scratch. This led to significant development efforts, which in turn led to pretty hefty price tags associated with participation and implementation. The value proposition and the anticipated ROI (return on investment) were impressive for large companies that stood to gain significant returns from controlling maverick spending and drastically reducing the cost of processing a single purchase order. However, the start-up cost proved to be prohibitive for all but the largest of companies. We will discuss these business issues later in this chapter as we talk more specifically about Internet procurement.

Several ERP vendors launched similar online exchanges early on in the E-commerce push, but some did so by partnering with either Ariba or CommerceOne. Some have further developed their own E-commerce platforms and superseded these partnerships, making for a rather discontinuous progression. But as the technology matures, these pioneering market makers penetrate markets more broadly and the ERP vendors strive to participate more fully, we can only expect this to drive the price down to a level where more universal acceptance is possible.

The more vertical approach to trading exchanges attempts to manage many-to-many relationships in more fragmented markets, in more complex product industries. These independent trading exchanges offer strong domain expertise in vertical markets such as plastics, chemicals, automotive after-markets, and electronic components. The domain expertise provided by these vertical marketplaces naturally lend themselves more toward the procurement of direct goods used in the actual manufacture of products marketed and sold. However, these marketplaces are not limited only to components and hard goods, but may also facilitate the buying and selling of services, such as transportation.

The appeal of these vertical exchanges lies typically in the added value specific domain expertise can bring. Participants can search for product based not only on price and predefined business relationships, but also on product characteristics specific to the industry. There is also the added appeal of market aggregation. Market aggregation can be leveraged in two distinct ways. Some market makers take on the responsibility of negotiating lower prices by leveraging the collective volume of all their buyers. In this way, participants hope to get a lower price by ordering through a marketplace than they could command on their own. However, another aspect of market aggregation lies in the ability to comparison shop. By aggregating a large number of suppliers into a single marketplace, it is much easier for a buyer to compare prices and other factors, without performing searches and hopping from one independent Web site to another.

While both Ariba Supplier Network and CommerceOne's MarketSite.net may have started out as horizontal exchanges facilitating the marketing and purchase of MRO goods, both have since expanded to support selected vertical industries such as consumer packaged goods, pharmaceuticals, chemicals, financial services and energy and the public sector.

As a manufacturer developing or executing an E-commerce strategy, it is highly probable that you will be drawn to participate in one or more of these horizontal or vertical trading exchanges, and in fact "more" is more likely. Many will use a portfolio approach to participation, using different models for different business requirements. Regardless of which of these exchanges you choose to participate in, ultimately your sale or purchase through the exchange must be mirrored in your ERP system. As a result, it is far more important that your ERP provider be able to interoperate with this exchange than to potentially compete with it.

Another method of categorizing online trading exchanges is by looking at who operates them, and here we see three categories — those operated by independent exchange operators, those run by a consortium of potential competitors, and private trading exchanges.

An independent trading exchange is a B2B trading hub, operated by an autonomous company that matches buyers and sellers and facilitates the transactions associated with the commerce conducted between them. Both Ariba Supplier Network and CommerceOne's MarketSite.net are also examples of independent trading exchanges. The company operating the E-marketplace is a new breed of intermediary, one that never touches the goods that are bought and sold. As such, this new kind of intermediary is not there to absorb inventory or the risk associated in carrying it, and therefore must add a different kind of value to the exchange of goods. The value these market makers have on the potential of offering includes the domain expertise and market aggregation mentioned above, as well as the ability to offer dynamic pricing models that can only be achieved through the introduction of many-to-many relationships. These dynamic pricing models include auctions, where a product is offered up to the highest bidder, and reverse auctions, where buyers attempt to set their own price.

As the concept was first introduced, expectations of acceptance and participation were huge. Industry experts saw this as the vehicle most likely to transport E-commerce into ubiquity. Market predictions of commerce to be conducted over these exchanges varied wildly among these industry experts, but they were in the billions or trillions of dollars. This business model has indeed proved viable for several vertical and horizontal marketplaces but the early market predictions have proved to be grossly overstated, although many industry observers see this simply as postponement of the inevitable.

The second form of exchange that has also not met early market predictions is the consortium-based trading model. These are sometimes referred to as industry-sponsored marketplaces. The concept here is that a consortium of potential competitors would collaborate to make a vertical market more efficient by standardizing processes and industry data. The other anticipated benefit was to provide services to individual companies that, if operating in isolation, would otherwise find them prohibitively expensive. In this case, the concept was to aggregate demand for goods or services to reduce cost and improve efficiency.

Covisint is probably the most widely publicized of these consortiums. Formed in February 2000 by Daimler Chrysler, Ford Motor Company, and General Motors, its purpose was to bring together individual E-business initiatives to avoid burdening common suppliers with the need to interact with redundant but different proprietary systems. Executives from these three companies formed a planning team and a new organization grew out of this collaboration, headquartered in Southfield, Michigan. By May of the same year, Renault S.A. of France and Nissan of Japan had joined and the new company was known as Covisint. By December, it had reached definitive agreements with technology partners CommerceOne and Oracle and it became a legal entity. The organization was formed as a multimember joint venture known officially as Covisint, L.L.C. — with the seven companies mentioned above as members.

Not all consortium-based trading exchanges come about in this way. The Nistevo Collaborative Logistic Network is a trading exchange that allows shippers and carriers to collaborate in order to share transportation capacity. Subscribers to this E-marketplace are consumer packaged goods companies, including Nabisco, Kellogg's, General Mills, Hormel Foods and Land O' Lakes, all of which share and compete for shelf space in your neighborhood grocery stores. Yet this consortium did not result from executives at these companies seeking to collaborate with competitors. Instead, it resulted from the vision of Nistevo Corporation's founder Kevin Lynch, who conceived of the idea back in 1997 and offered these companies a way to improve efficiencies and lower their transportation costs in 2001. So, by the strictest definition, this is not truly a collaborative exchange, as it is not run by a group of competitors seeking to collaborate, but rather it is a unique version of an independent trading exchange.

Participating companies submit routes and information detailing contracts with shippers. Nistevo matches them up and notifies potential partners. Once agreements have been worked out, shipments are scheduled, which can potentially have competing products delivered in the same truck. Shippers and logistics providers are made more efficient and this results in savings to the customer. The added benefit is that Nistevo is able to connect transportation providers with these consumer packaged goods manufacturers regardless of the data exchange standards they use, thereby addressing the issue that led to the formation of Covisint and other consortium-based exchanges.

The purpose is well intentioned, but the concept of collaborative competition has not been wildly successful. The struggle, as with any type of standardization effort, is over consensus, which is the reason E-marketplaces such as the Nistevo Collaborative Logistic Network have a far greater chance of success. Operated by an independent company that is willing and able to serve as a gateway between a variety of standards, much of the requirement for consensus is removed.

The third category of exchange is the private trading exchange. This is a single company's trading interface to the world via the Internet. We could argue that the private trading exchange is not exactly an intermediated business model. However, it quickly moves beyond the concept of direct links between companies, or even a model of a one-to-many relationship. It is based on the same concept of the many-to-many relationship model employed by other trading exchanges. Many start simply as an information portal, similar to Belden Brick's customer Extranet that was described in Chapter 4. Today Belden simply allows independent distributors to communicate through their list server, over their Extranet. But consider the possibility of a next step that would bring E-commerce capabilities to these distributors. This would provide Belden the vehicle to operate a private trading exchange, through which these distributors could directly conduct business. The extent to which Belden might provide services to facilitate these transactions might vary extensively. Currently they only provide a vehicle for communication, but in the future, who knows what future market conditions might prompt them to provide. They may continue to serve only as a clearinghouse of information, or they may decide to play a more active role in facilitating the transaction itself, either electronically or physically, or both.

Other private trading exchanges may result from the continued growth and diversification of large, multidivisional, multiorganizational companies, and the resultant need to simplify and enhance their own processes. Even in simply facilitating E-commerce with its own customers and/or suppliers, requirements can easily grow to demand control of the same complexities of an independent trading exchange and its many-to-many business model. While there are certainly numerous factors involved, the decision to create a private trading exchange or to participate in an independent or consortium-based exchange might boil down to a single question: Do you want to allow other companies to secure profits from offering your products through their exchange, or would you prefer to increase your profits from the converse — offering other companies' products through your own exchange? While on the surface this appears to be a simple question, you need to weigh very carefully the cost of establishing or purchasing your own E-commerce platform with the value proposition you can bring to your customers in doing so. What complementary products or services are you able to bring to your customer that increase the value proposition you can present? In ordering a computer from Dell, the consumer can add software from Microsoft or other vendors and printers from Epson or Hewlett Packard. They can finance the transaction over an extended period of time. Yet they never have to leave Dell's Web site or make additional phone calls. The intercompany operations between these different enterprises are transparent to the consumer. Yet few companies have the market dominance of Dell Computer. The struggle with the private trading exchange tends to be with establishing the necessary levels of trust between all trading partners and handling multiple points of integration.

The bottom line of private trading exchanges, as well as other forms of E-commerce, is the understanding that relationships based on trading partners absorbing additional costs will be a hard sell. Chances are you will only be successful requiring these partners to take on this burden of time and money if you enjoy the market dominance of a Dell Computer, or represent the primary source of revenue for your suppliers or you can somehow provide a new or increased source of profits to your customers. Therefore, the more variation you can accept in terms of data exchange standards and the less specific processes you require, the easier it will be to gain acceptance.

E-COMMERCE PLATFORM REQUIREMENTS

One clear advantage of an independent trading exchange is that it is hosted by its market maker, thereby eliminating much of the setup and maintenance cost of the infrastructure required to support the electronic relationships between participants. Those companies seeking to establish a private trading exchange, or a group looking to creating a collaborative consortium need to consider the E-commerce platform required to support such an endeavor. While early exchange operators like Ariba and CommerceOne had to start from scratch, today many of the elements required to support these trading hubs can be purchased from software suppliers.

This is a relatively new software category and the market is comparatively immature. While other categories of software, like ERP, have become commodities, with most vendors providing at least a standard subset of functionality, E-commerce platform providers are a more varied lot. It is gratifying to know that many of the entrants into this new space are well-established companies and best-of-breed providers in certain of the elements required from an E-commerce platform. However, it is still not unreasonable to anticipate the natural selection process, which is bound to occur as the market matures. This means competitive shakeouts, acquisitions and mergers. As a result, you will have to be very clear on what you expect to gain from establishing a private or consortium trading exchange in order to match your expectations carefully with the functionality currently provided by the potential software vendors. It may also mean that you will have to prioritize what functionality you require now and what you may be able to postpone until future releases of the software are available. However, this prioritization need not be viewed negatively, but simply as a smart and pragmatic way to establish a private or collaborative trading exchange. It is unrealistic to assume a total business process reengineering, particularly with a diverse set of participants. Therefore it is appropriate to be realistic in choosing which business processes to automate first.

However, there is certain feature functionality that is basic to whatever you hope to accomplish with a private trading exchange. First and foremost is the infrastructure required to provide the necessary connectivity, security and administration of the trading hub. This translates to the provision of both users and machines interfacing with the E-marketplace. Since it will be a many-to-many business model that is represented, user profiles, permissions and varying levels of security are a necessity, and you will need to manage the network and the systems connected to the network. If this seems a daunting task, do not discount the possibility of outsourcing it. Many of the E-commerce platform providers will also offer to host the trading exchange for you, therefore relieving you of the burden of managing this foundation.

The lifeblood of the E-marketplace, however, lies in its ability to manage the commerce, which is transacted over the exchange. This includes the ability to match buyers and sellers to the advantage of both. For companies to willingly participate in this private exchange, some value must be offered. Sellers find buyers, buyers find suppliers or some other added value is derived, such as the communication for collaboration that Belden Brick provided its distributors simply through its Extranet of E-information. The ability to take an order is, of course, the minimum requirement, but trading exchanges are further enhanced with dynamic pricing, Request for Quote (RFQ) and Request for Proposal (RFP) capabilities, as well as the ability to quote, configure products, create and negotiate contracts and to monitor status.

Beyond these basics, there are other added-value services that can be offered to attract participants to a trading hub. These may be related to information shared throughout the trading hub. The business community may benefit from content in the form of industry or process-specific information. Depending on your goals, you may seek an E-commerce platform that allows only the market maker to publish this content, but may allow different participants to subscribe to it. Or you may require the ability for any of the participants to securely create, publish and subscribe to either structured (e.g., catalogs) or unstructured content.

Personalization may be an attractive feature or it may be an absolute must. Negotiated pricing, preferred customer relationships and customized catalogs all require a mechanism for recognizing and acknowledging individual customers. This may be as simple as overriding standard catalog prices with contractually pre-negotiated discounts, or as sophisticated as dynamically tailoring product offerings based on the recognition of buying patterns. Consider these options carefully, particularly where customer relationships have been historically cultivated by personalized human contact.

Look for additional value that can be added through intelligence gathering and workflow automation. A potential goldmine of data can be captured through an exchange. Without providing the ability to analyze this data and present the results in a meaningful way, it can lie unmined. The implementation of simple workflow capabilities can serve to provide an added level of process automation. This can also be supplemented with a good rules engine, as introduced in Chapter 4. The ability to define business rules, monitor events, trigger alerts and workflow events are important considerations in enhancing communication between trading partners. And communication is the underlying foundation of collaboration, which is in turn the underlying goal of any trading hub.

And finally, consider the fact that E-commerce will increasingly involve partnering beyond just the buyer and the seller. To automate and optimize the buying and selling processes from end to end, other affiliated companies will play an amplified role. These companies may be banks, credit card processors, third-party logistics providers or any number of other businesses that potentially add value to the relationship between the buyer and the seller. The business models tying all these partners together will become more virtual. The ultimate test of the E-commerce platform will be its ability to make these interactions appear seamless, not only to all participating entities, but also, most importantly, to the customer. This seamlessness will be characterized by short implementation times, which are the result of the exchange's ability to accept a variety of data and exchange formats and the partners' ability to plug easily into the infrastructure.

A SHAKY BEGINNING

As long as 15 years ago predictions were that EDI would be a universally accepted, and indeed preferred, method of doing business by now. Yet implementation is still far from universal. In a similar fashion, anticipated revenues generated through online trading exchanges have fallen far short of early predictions. Sounds like history repeating itself? Yes and no. Yes, as in implementing EDI, many companies underestimated the time and resources necessary to provide a complete E-commerce platform or infrastructure necessary to make these exchanges run. Early exchanges did a good job of matching buyers and sellers but left a lot of the details untouched. They also underestimated the scope of the integration effort to plug into them. And participants overestimated their ability to effectively interoperate within an integrated business community. However, there were also other factors involved, which were of an entirely different nature.

One factor that never came into play with EDI was the low barrier to entry in establishing an electronic marketplace, in combination with venture capitalist zeal for the concept. When the barriers to entry are low, and money is available, you will get many new entrants quickly. When this is the case, simple laws of economics regarding supply and demand kick in. Scores of independent trading exchanges sprang up, virtually overnight, making competition fierce and margins low. When large numbers of exchanges emerge virtually overnight, no single one is able to gain a dominant share.

Trading exchanges also miscalculated how much customers were willing to pay for which services performed. This was compounded by management teams and backers with little experience in the markets they served. As exchanges struggled to capture much-needed market share, many of the early entrants found themselves redefining business strategies and business models, merging with competitors, or simply going out of business.

And finally, the purpose of EDI was never in question. The goal was always to streamline and automate processes ultimately to save time and money. Reduced cost and improved efficiencies were also a goal of E-commerce, but there was far less clarity in terms of the ultimate goal of implementing electronic trading exchanges. This begs the question: What is the true value of an E-marketplace?

THE VALUE E-MARKETPLACES BRING

Why do companies participate in independent trading exchanges? What do they intend to accomplish by implementing a private trading exchange or by collaborating with a consortium? And, finally, does the value proposition offered by these E-commerce initiatives match the desired results?

Because of the very nature of an exchange, there are two perspectives from which to evaluate the potential benefits it can bring. There are those companies that seek to improve customer-facing processes. These are the companies seeking to sell their products over the trading hub. Then there are those companies that seek to improve supplier-facing processes. Obviously, these are the enterprises seeking to purchase materials through the hub.

Those companies that seek to improve customer-facing processes are either looking to streamline and automate these processes, or they hope to gain new revenue. Those that entertain the possibility of participating in an independent trading exchange that already has a well-established presence and a solid cast of characters typically hope to gain new customers from their entrance. Those that seek to initiate a private trading exchange are more likely looking to increase business from existing customers by offering added value to these customers and locking in the relationship. Those seeking to collaborate with competitors are typically seeking to streamline and standardize processes for the purpose of improved efficiency and potential cost savings for themselves and possibly their customers as well. Improvement of supplier-facing processes is almost always characterized by cost savings or efficiency improvements, or both.

It is not surprising that early marketplaces did not take off as quickly as predicted. The early exchange operators made market aggregation the emphatic and sometimes exclusive value proposition. The promise of making the purchasing power of the community available to each individual was compelling. However, the cost of entry generally precluded the small to medium-sized business that stood to gain the most from this leverage point from participating. These early trading hubs rolled out industry news as a display of their domain expertise, and auctions as their key differentiator. Auctions were new. And reverse auctions were touted as the cure for the age-old problem of excess and obsolete inventory. Stories of price discovery and spend aggregation abounded. A typical business scenario described by marketers was an automated supplier search in order to satisfy demand for goods or services. While this might play well in the search for a new supplier or for a spot buy, the market makers went too far in suggesting that this automated search be conducted for each and every purchase. This scenario made customer loyalty obsolete and trivialized buyer–seller relationships. While it may have appealed to those caught up in the new business models of the pure dot-com, those with experience in the real world of traditional buying and selling values viewed it with suspicion, and rightfully so.

Pragmatic companies were not looking to replace experienced buyers; they were looking for ways to make them more efficient. And that meant freeing them from the dog-work of follow-up and status monitoring, so they would have more time to perform the analysis and negotiation necessary to build solid relationships with key suppliers. And, therefore, it is not surprising to learn that market aggregation, industry news and auctions were not on the top of the list of what they were looking for. What is on the top of the list, without a doubt, is the ability to share information efficiently among the trading community. This reflects the fact that buyers in companies with only traditional means of communication spend far too much time chasing vendors for information regarding late or missed shipments or documentation, when they could be more proactively managing supplier relationships. And customer service representatives and sales personnel are similarly inefficient in providing customers up-to-date status information. Yet the type of information that can be efficiently shared need not stop at order status, but can extend to engineering or quality documentation, as well as advanced planning and forecasting information.

Also considered important is the streamlining and automation of the buying and selling processes, which are at the heart of commerce. This type of automation is the primary source of savings in terms of reducing the cost of the transactions themselves, which must be distinguished from reducing the product cost of the purchase. However, many companies do regard reduced procurement cost as another viable reason for considering participation in an exchange of some sort. This is the one direct benefit that is possible through market aggregation, using the buying power of the consolidated exchange to the benefit of each participant. The two cost savings together have the potential of a considerable reduction in total operating costs.

Some enterprises also view trading exchanges as useful in identifying new suppliers and new products. On the sales side, they also see some benefit in improving their reach into their market, particularly through the use of an established independent trading exchange, just as the retail shop will seek space in the busiest of malls.

While seldom will a company participate in an exchange for the sole purpose of producing sales of excess inventory, this can be another useful benefit if the trading exchange can support auctions and reverse auctions.

EVALUATING TRADING EXCHANGES

The dot-com deathwatch has caused most companies to give far more careful consideration before jumping into any B2B E-marketplace. This caution is appropriate but no more difficult than the analysis of any key business case. And when considering an independent trading exchange, in many ways it is more easily accomplished than the analysis of other new sales channels. When evaluating a potential trading exchange, there are three important criteria to consider: the functionality offered, the credentials it carries and the current and anticipated health of the exchange.

When evaluating the functionality offered by the marketplace, a good starting point is simply to ask the question, "What do they do?" Is this a communication hub whose exclusive purpose is matching buyers and sellers, but no transactions are conducted online over the hub? If this is the case, the exchange is simply acting as a facilitator and perhaps an information provider for industry-related news and information. One might argue that in this case the marketplace might not truly be considered a trading exchange. However, since these types of exchanges may eventually evolve into trading hubs where legally binding transactions are consummated, it may still be appropriate to include this type of organization in your considerations.

If in fact the exchange goes beyond matching buyers and sellers and these transactions are conducted online, the next logical question to ask is whether these transactions are integrated into participants' back-office applications. If so, the specific requirements for this integration and the level of effort required to achieve it is a very significant point of consideration.

Given the varying levels of marketplace maturity, it is also important to determine which, if any, value-added services are offered beyond the basics of transactions. These might include trade financing, logistics, provision of a clearinghouse for payment and settlement, as well as electronic invoicing and statement generation, marketing services, credit evaluations or reporting and insurance. It is also appropriate to determine the level of reporting provided as a standard service and how that might be extended through additional software services or data mining. An enormous amount of statistical and demographically oriented data may be captured through the electronic buying and selling process, but don't simply assume this information will automatically be made available to you. And, finally, simply ask what other unique services or complementary products might be available to you or your customers or suppliers through operation within the exchange.

The second criterion for selection should be the expertise or credentials carried by the organization operating the exchange. What kind of domain expertise does it offer? Is it specific to a business process, such as procurement, or product or industry specific? How well does the exchange know your business, including industry-specific processes? When evaluating these considerations, it is important to consider this from several perspectives. Obviously before entering into a business partnership of this scope and nature, it is naturally expected that you will check out the management team. However, it is not always so customary also to check out the expertise of the people you will be interacting with on a more routine basis. Does the exchange offer a strong customer service staff with domain expertise and industry knowledge?

Look carefully at the methodology used for implementation and determine a realistic level of expectations for time and effort for deployment. Is the implementation fast and hassle-free? Does it include installation and training for any software that must be resident at your location. Determine if the exchange was designed with the buyer in mind or the supplier, or both.

You may also determine if there are strong ties to an industry organization, preferably one that is viewed as the definitive source of industry information, knowledge and innovation. But you will also want to determine if the exchange operates neutrally or is driven by the influence of a significant equity stakeholder. Some trading exchanges seek third-party partnerships with established brick and mortar businesses within an industry in order to draw others to it. If this is the case, they seek the most dominant players in the industry. For example, the electronics components industry serves many different vertical markets including high-tech electronics, consumer electronics, automotive and consumer durables. This industry is characterized by highly segmented supply chains including component manufacturers and distributors, board and system level contract manufacturers, system distributors and re-sellers. There are large and influential players in each segment and many of these supply chains operate under the auspices of a brand owner such as Motorola, Compaq or Cisco. If in fact one or more of these influential companies has taken a significant equity position in the trading exchange in order to use it as a sourcing or distribution source, the exchange's neutrality may be somewhat compromised. However, each individual company must determine if this is a deterrent or an attraction to participation.

Once you have established this level of credentials, you can start to ask the harder questions. If the exchange touts itself as having a global reach, try to determine the geographical dispersion of participants. It may claim to be a worldwide exchange, but if 95% of its participants are located in the United States, this fact speaks louder than the marketing claims. However, if the trading exchange does successfully operate worldwide, investigate how this is done. If the trading exchange plays more of a role than simply matching buyer and seller and electronically facilitating a transaction you make directly with your trading partner, there may be currency and other financial restrictions. What currencies can be exchanged? How are currency conversions handled? What international trade restrictions are you required, by law, to operate under? Are these recognized by the marketplace and enforced?

Perhaps the most frequently overlooked question to ask concerns what happens when something goes wrong. Does the marketplace offer a service-level agreement to participants, and what impact does that have directly on you as either a buyer or seller? If you are the seller, are you able to fulfill orders at this level? What happens when there is a disputed transaction? Exactly how much liability does the market maker absorb and how much is passed on to you? And what commitments will the exchange operator make to you in terms of access to the exchange? Most claim to operate 24×7, but there may be some level of expected and acceptable downtime. You should ask the tough questions. How many hours of unscheduled downtime did your marketplace experience last month and last quarter? What were the reasons for this downtime? What kind of business continuity plan do you have in the event of a major man-made or natural disaster?

The third and final criterion should come as no surprise to any businessperson, as it deals with the overall health of the trading exchange as a business enterprise. In the beginning of this chapter I made reference to the fact that the success of some trading exchanges, particularly horizontal exchanges, hinges on the momentum that can be achieved. Therefore, first and foremost, you need to determine if the B2B marketplace has sufficient volume to sustain its current position and to grow. Volume can be measured both by the number of buyers and sellers and also by the number of transactions processed daily. Of course, for those exchanges that simply match buyers and sellers and do little else to consummate the transaction, you may be tempted to re-define the criterion to measure the number of buyer to seller matches. However, use this statistic, if available, with caution. The number of matches, without the number of transactions that are generated from them can be very misleading. If a lead source generated hundreds of leads but none of them turned into real business, would you consider the source of these leads valuable?

Once you have determined these preliminary statistics, consider the average size of the transaction, as compared to your average order size. There may not be a direct comparison between the two. If previously you sold to distributors and this marketplace connects directly to individual retail outlets, you may be comparing apples and oranges. However, do not overlook the significance of the discrepancy. If you were fulfilling truckload orders and now the orders are for pallets or individual piece parts, the success you enjoy through the trading exchange will be short-lived unless your back-end fulfillment process can adapt.

Also consider the total revenues that are transacted through the exchange. These numbers are best compared against the total industry revenue. What may first appear to be very significant revenue may in fact be a rather insignificant percentage of market share. If this is the case, the numbers may either be telling you that you are looking at the wrong exchange or that this market is not quite ready for this as a viable sales channel. If the latter is true, you need to weigh the value of first mover advantage and make a strategic decision.

It is just as important to determine how the revenue is achieved as it is to determine how much revenue there is. Typically, revenue can be achieved through a B2B marketplace either through advertising or through fees associated with doing business through the trading hub. If most of the revenue is driven through advertising, this is not a strong indicator that real industry revenue is flowing through the exchange. There are different business models for generating non-advertising revenue. Some fees are subscription based, which are similar to software license fees, and some are transaction based. Many exchanges use a mix of these. It is not uncommon for a single trading exchange to have a monthly subscription fee for its buy-side application, and offer its sell-side application free of charge. However, this free deployment will usually be very basic and is typically not connected to any back-office applications. The same exchange operator may offer a more robust application that can be connected to a back-end system for some sort of fee. These fees may either be flat fees, regardless of volume of business conducted, or they may be somehow transaction based. The fee paid per transaction will generally decrease as volume increases.

In addition, the market makers may also have other sources of revenue, including professional services fees associated with implementation, data mining or reporting. In general, it pays to ask your potential exchange operator for a breakdown of its revenue in terms of all these categories, including software licenses, transaction based fees, data mining and other services.

And finally, you will want to determine the rate of growth that has been achieved by the exchange and, although somewhat subjective, determine the potential for sustained growth.

THE BUY SIDE OF E-COMMERCE

The buy side of E-commerce is often a compelling starting point for companies because it holds the potential of carrying a high return on investment, and it avoids a lot of the political landmines, such as channel conflict, that the sell side can encounter. Your approach to the buy side can be very different depending on whether you initially target direct or indirect items. Direct items are used in the manufacturing of a company's finished products and indirect items are everything else. Most direct procurement initiatives to date have evolved from Electronic Data Interchange (EDI) programs. Many times discussion of Internet procurement of indirect items will be confined to MRO items, but in fact this general category also includes items such as travel and expense, contract personnel and advertising. Early buy-side E-commerce initiatives focused primarily on MRO purchases mainly due to the compelling returns from improved compliance, improved negotiations and reduced process times. While restricting your E-commerce initiatives to MRO may prevent you from gaining the benefits of controlling all your indirect spending, it is easy to see why this happens because these other spending categories, particularly travel and expenses, may require different solutions. Because travel and expense can be entirely separate and distinct from the purchase of other goods and services, this section will simply recognize it as a separate spending category but will leave this discussion to the travel experts.

THE BENEFITS OF THE BUY SIDE

The potential benefits of buy-side E-commerce are substantial. MRO-oriented initiatives have been the most popular implementations, not only because of the potential payback, but because they represent an opportunity to provide a beneficial service to so many people within the organization. The returns center primarily on cost savings and efficiency gains, with the most obvious the lowered cost of administration of requisitions and purchase orders.

Lower Purchase Administration Cost

While different industry experts will offer different average costs for the processing of an indirect purchase order, they seem to converge around the broad range of $50 to $200, although the highest average cost I have heard a company admit was somewhere around $350 per purchase order. However, this was in Helsinki, and I can only hope this outrageous estimate was a result of a mistake in currency conversions. However, even the lowest averages that experts quote are a bit frightening considering the cost of processing a purchase order for a $10 wrench can be the same as processing one for a $4000 laptop computer. When the processing cost exceeds the cost of the purchased item, that alone should provide enough incentive to companies to seek alternative, less expensive procedures. And, indeed, lowered administrative costs for procurement are typically cited as the number one advantage to Internet Procurement. Applications that specifically target indirect purchases automate much of this process and save a significant amount of the employees' time. Early implementations of these types of projects can bring the cost of this administration down to $10 to $15.

Control Maverick Spending

The second most sought-after gain is improved compliance with corporate contracts and standards. Controlling "maverick spending" is a popular theme when E-procurement software vendors talk about benefits of implementing their applications. Maverick spending occurs when purchases do not leverage corporate contracts and volumes. The purchasing department will negotiate contracts with suppliers for substantial discounts. When these agreements are ignored, the opportunity for savings is lost. Most maverick spending is neither a malicious nor a deliberate attempt by employees to thwart corporate policy, but instead results either from a lack of knowledge of these agreements or from convenience. Visiting the local office supply, hardware or computer store is simply faster and easier than going through a standard requisition and purchase cycle. An objective of E-procurement is therefore to make the requisitioning process as simple and nearly as fast as a trip to a nearby store. Again statistics vary but the experts tend to estimate that prior to implementation of E-procurement 27 to 35% of indirect purchases are noncompliant.

Shortened Purchase Cycles

The corollary to reducing the cost of administration is the reduction in the time it takes to complete the full cycle of requisitioning through fulfillment. For indirect purchases this works hand in hand with the control of maverick spending. By reducing the time it takes to satisfy a requirement, you take away one of the major reasons for noncompliance. For direct goods, this can also have an added benefit of reducing inventory. By shortening the procurement lead time, less inventory needs to be kept on hand to cover the requirements within that lead time.

Lower Price of Purchased Items

There is a second opportunity for lower costs, and this lies in the cost of the purchased item itself. By aggregating spending on like items and improving the data gathering and reporting, you can enhance the negotiation process and gain additional leverage in the negotiation with suppliers. Even a very small increase in discount percentages can add up to very large savings depending on the total purchase amounts.

The savings from the buy side of E-commerce should be expected to be cumulative, over time. While reductions in the cost of administration can appear very significant and immediate, some of these savings can be soft savings and may not be directly measurable unless employees are actually eliminated. This is not necessarily the case. In many instances, these personnel are instead re-deployed. In fact, one side benefit to the purchasing department is that procurement professionals can be freed from pushing reams of paper orders in order to focus instead on more strategic tasks like managing relationships and negotiating discounts. Savings in terms of lowered product costs may not be so immediate, since data must be gathered and aggregate spending reported and analyzed before discounts may be renegotiated. However, even the most conservative estimates indicate that implementing E-procurement, even if it simply targets MRO purchases, yields relatively rapid payback and a high rate of return.

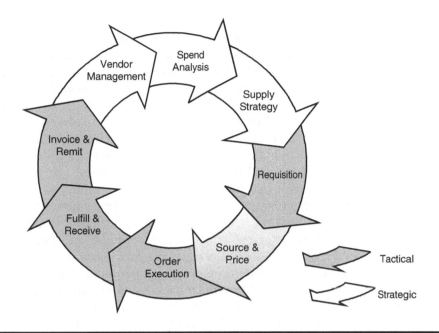

Figure 5.3 Procurement life cycle.

THE PURCHASING PROCESS

Figure 5.3 depicts the various processes associated with procurement, whether it is done using traditional methods or is electronically enabled. These processes can be characterized as either strategic or tactical. Those shown in the gray arrows are the tactical processes that have been automated by E-procurement, while the white arrows indicate the more strategic activities necessary for E-sourcing. E-sourcing is often also referred to as Strategic Sourcing. Direct and indirect procurement follows the same basic life cycle, but at certain steps in the process the characteristics of each subprocess can vary significantly.

Spend Analysis

If we start at the top of Figure 5.3 with spend analysis and work clockwise around, you will find that the entire process is a closed loop. The spend analysis attempts to determine how much money you are spending, what you are spending it on, and with whom. Breaking down your spending will drive much deeper than simply distinguishing between direct and indirect. Every spending category will have a unique set of physical supply issues, as well as strategic sourcing requirements. Doing a spend analysis will determine which spend categories can yield the highest rate of returns. If you are a plastics manufacturer doing injection molding, how you reduce the cost of the purchasing process and the cost of the raw materials will differ significantly from a discrete manufacturer with large bills of material (BOMs) containing many standard components. You obviously don't need to perform any kind of sophisticated analysis to tell you this. And simple inventory accounting should easily determine how much of your aggregated spending is on direct and indirect materials. This is information your ERP system should be able to give you fairly easily. If it does not, reporting can be supplemented with a good report writer.

However, can you easily determine how much of the money you spend is based on negotiated contracts vs. spot buys? Or how many of those spot buys could have been purchased under a negotiated contract? And can you determine how effective your existing contracts and negotiated prices are in reducing the cost of purchased product? How do you know if the same item is ordered by the engineering department using one part number and the maintenance department using another, and manufacturing using yet another? You may have an identical part purchased from a different supplier and stocked in a different location under a different part number. You may eventually see the problem for direct materials, using standard obsolescence reporting through your ERP system, but what if these stockable items are indirect, perhaps equipment spares?

Category management is the key output spend analysis and will be input to the supply strategy process. Category management creates a hierarchy under which to aggregate spending. Unfortunately, there is no single scheme for categorizing materials, although the United Nations Standard Product and Service Code (UN/SPC) does provide one that could be called dominant and several software companies that provide analytical tools base their suite of products and services on it. However, which scheme you use is less important than simply using a scheme — any scheme.

However, which software tools you might have available to analyze the results will vary significantly depending on what kinds of materials you are categorizing. There are software vendors that specialize in content cleansing tools and services, particularly for non-stocked items. Component Supplier Management software manages and reduces part duplication often found where there are large BOMs of standard components. Redundant parts create excess inventory and inflate purchasing costs. For a manufacturer that uses standard components, and that seeks to collaborate within a supply chain, addressing this problem is key. If you need to identify contract leakage or supplier data quality issues, this is another whole series of software vendors and products. Dun & Bradstreet is the market leader in this space, but there are numerous other niche vendors that provide data rationalization and analytic tools specific to spending analysis. Generally, these software products use your accounts payable system as their primary source of data. If you want to take the results of your spending analysis and feed them to your ongoing vendor management efforts, then an investment in business intelligence probably makes the most sense.

As you can see, there is no one particular path you can take to reap the benefits of spending analysis. However, without truly understanding how money is being spent corporate-wide, you may be blind to opportunities for improvement.

Supply Strategy

Supply strategy takes input from your spending analysis, and also from other sources, and attempts to address issues such as core competencies, make vs. buy strategies and supplier rationalization. Which categories do you focus on? How much should you consolidate your supplier base? Should you reduce it to the point where you have a sole source for each item or for each category? This process determines which vendors you will consider, the processes you will use to evaluate them, as well as the criteria you will use to measure them in the evaluation stage and later as you perform ongoing vendor management.

Setting supply strategy is not a mechanical process that can be automated and therefore there is no single piece of technology that you can point to that will guarantee you accomplish it. Yet there is technology that you will apply to other purchasing processes that can potentially have either a hidden or a very visible impact on your strategy. E-procurement, auctions and reverse auctions, and Internet-based trading exchanges all serve to smooth the buying process and reduce prices, but they can actually damage the relationships you have with your suppliers. The decision to implement buy-side E-commerce will require you to consider alternative approaches carefully, and these approaches will undoubtedly vary by category. Which products will you open source, treating each as a spot buy and possibly negotiating each purchase? Which products or categories will you manage with mid- to long-term contracts that are renegotiated periodically? This is where the RFP or RFQ or bidding and negotiation process is defined, if appropriate. Or will you decide to have a very close relationship with fewer suppliers? These decisions should not be driven by the lure of the latest technology available, but by consciously defining your supply strategy in the greater context of your overall business strategy. Then use this strategy to determine the direction your E-commerce initiatives should take. In other words, don't let the tail wag the dog.

Requisition

The requisition phase of the procurement process is where requirements are generated. It is at this point that there is a significant divergence between direct and indirect purchases. All ERP systems today can automatically generate requisitions for direct materials using traditional Material Requirements Planning (MRP). However, not all manufacturing environments necessarily lend themselves to MRP planning techniques. Whether these environments use KANBAN techniques or simple reorder point methodologies, it is safe to assume that the generation of these requirements can be automatically created through your ERP system.

Except for indirect materials that are kept in stock and generally managed on a reorder point basis, requisitions for indirect materials are more likely created manually on an *ad hoc* basis. It is during this phase where buy-side E-commerce can begin to improve operational efficiencies. In the traditional procurement scenario shown in Figure 5.4, individual buyers complete a purchase requisition (req), then route it through approval channels. Purchasing approval chains can consist of a succession of staff members who have the authority to approve or reject orders. The sequence of approvals may require multiple paper documents and numerous follow-up phone calls. Most ERP systems today support the generation of the purchase requisition, but they vary to a great degree in their ability to manage and automate the approval process.

One of the objectives of E-procurement is to streamline this process dramatically by putting the requisitioning process directly in the hands of the individual buyers. See Figure 5.5. Theoretically this could be accomplished using the requisitioning process in your ERP system, but there are a few caveats to this. First of all, decentralizing procurement requisitioning, from a systems standpoint, expands the user base from a limited number of purchasing agents to hundreds, thousands or even tens of thousands of employees. This has implications for your ERP system both in terms of performance and cost. Most companies pay for ongoing maintenance of their systems either based on the number of users or on the size of the computer on which it runs. By unleashing a potentially large number of additional users on the system, this could potentially cause the ongoing cost to rise appreciably.

If most of your office employees already have access to the ERP system, this may have a negligible effect. Could your ERP system be enhanced to handle this if this were the case? Several issues must be considered. The first is ease of use. Currently, your purchasing agents are well trained to use the procurement functions of the system. How much training would each employee who is a potential buyer require? Is the requisitioning process truly intuitive? And what about the requisition approval process?

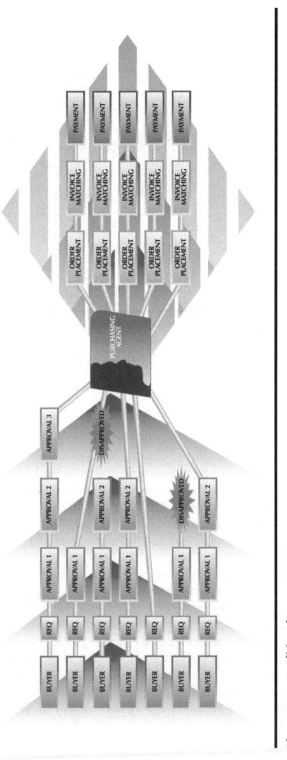

Figure 5.4 Traditional procurement.

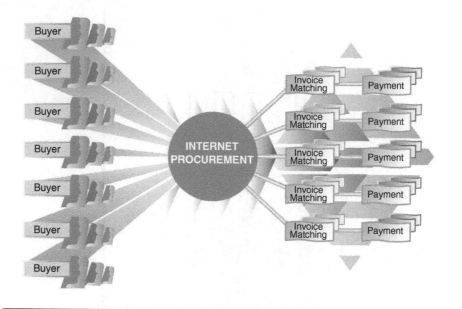

Figure 5.5 Internet procurement.

Is it fully automated? The best way to automate the sequence of approvals is through workflow technology.

Source and Price

The source and price process in Figure 5.3 is the transition phase between strategy setting and execution. It can be somewhat strategic in that it is the process through which relationships are established and contracts are negotiated. Of course, once these agreements and relationships have been established, the process becomes tactical.

The Internet has long been a source of information to purchasing professionals researching possible new vendors. Standard key word search engines can identify potential sources of supply. E-marketplaces, particularly those focused on specific vertical industries took that concept and developed it further. But whether you choose to participate in these collaborative or independent trading exchanges, or whether you prefer to establish individual B2B relationships through a private exchange or through specific relationships, E-procurement provides you with additional choices in executing the supply strategy you have established.

In addition, point solutions have become available that are aimed specifically at E-sourcing. These applications provide critical decision-support capabilities and facilitate the determination of suppliers, allocations of contracts and volume and management of supply risk. Their prime objective is to bring discipline to the decision-making process while at the same time reducing the time it takes to reach well-informed decisions. For most companies the most pervasive analytical tool used by purchasing managers is the spreadsheet. Yet spreadsheets are inadequate for analyzing the complexities that arise when supplier selection goes well beyond shopping for a combination of product features and the lowest price. By providing a combination of content, pre-negotiation tools and analytics, E-sourcing solutions can form the building blocks of effective supplier selection, allocation and negotiation. By bringing an automated methodology to the table, companies are able to consider a larger number of competitors without dragging out the process.

These E-sourcing applications can potentially create templates for RFIs/RFQs and provide a repository for detailed specifications, market information and supplier capabilities. Using these types of tools, the potential buyer can define the attributes required, along with weighting factors and the terms and conditions desired in order to support structured negotiation and decision support. They can facilitate the process of ranking the responses automatically against your criteria and subsequently support the submittal of offers and counteroffers. They can create an environment within which competitive bidding and negotiation activities can occur online, which can also be merged with off-line negotiations.

While automating the traditional RFI/RFQ process is one method of E-sourcing, auctions provide an alternative approach. Auctions are certainly not new. They date back to very early civilizations where the most common items auctioned were spoils of war. However, conducting auctions over the Internet in a B2B setting is a relatively new concept. Until recently, auctions were based on price-only, and therefore had limited applicability where selecting the lowest or the highest bid wasn't always the best decision. Making supply chain purchases tends to be more complex than this. Not only must you consider price, but also a myriad of other considerations including delivery performance, quality, lead time and transportation costs, to name just a few. Yet auctions do provide a viable alternative in certain cases and, as a result, auctions today are taking on new dimensions.

The most traditional form of auction is one in which the price ascends over time. There is one seller and many buyers. These may be open auctions or closed, where participants are welcome by invitation only. This has become an attractive means of selling off excess inventory, and is therefore referred to as a surplus auction. This, obviously, is the attraction for the seller. Buyers are attracted by the possibility of significant cost savings. The most simple of these auctions are based on price-only, but more sophisticated auction applications do exist that allow buyers or sellers to formally indicate nonprice aspects, allowing the introduction of purchase criteria such as warranty and service factors. Where other factors besides price are being considered, invitation-only or closed-bid auctions are more common.

The seller initiates ascending price auctions, but reverse auctions are initiated by the buyer. This has become the more prevalent mode of auction operation in a B2B environment. Reverse auctions operate in downward mode. There is one buyer and many sellers. The price goes down as the buyer receives bids.

These auctions can be run as a single event, or on a recurring basis or simply as a means of dynamic pricing. Single event auctions are typically run in cases where goods and services are purchased on a long-term contract. Recurring auctions are more appropriate for shorter-term contracts. Dynamic pricing is transactional in nature, as opposed to contractual. It is used in the auction of surplus goods and other instances where spot buys are appropriate. Dynamic price auctions are typically price-based exclusively.

Typically, these E-sourcing application vendors can also host the application, reducing the investment in technology infrastructure required from the participating companies. As a result of these hosted environments they can create online business communities, and immediate access to many, possibly hundreds of suppliers becomes available. This is particularly helpful in identifying possible candidates to consider in the sourcing and pricing phase.

Transaction Automation

The next three phases in the procurement life cycle, order execution, fulfillment and receiving, and invoice and remittance, comprise the primary transaction phase, which is automated through E-procurement applications. These applications eliminate much of the paperwork, speed processing and reduce the cost of transmission of business documents. In approaching this portion of the procurement cycle, it becomes more important to distinguish between the purchase of direct goods and that of indirect goods. Not only do business managers typically view direct materials as critical to their competitive advantage, while the majority of indirect goods are ancillary or supportive of the business, but also the actual procurement processes can be very different. As a result, it is quite probable that these functions will be managed not only with different approaches, but also with different applications. For this reason, this chapter devotes separate sections to each of these general categories.

Vendor Management

Most purchasing departments today have some formalized process for determining vendor performance, which is an important factor in overall supplier relationship management. It is for this reason that the procurement life cycle must be viewed as a continuum, where this performance analysis, together with a constant watch on spend analysis, provides ongoing input to supply strategy. Much of the data required for vendor management will be derived from your ERP system, or its back-office equivalent. Price and delivery performance can be derived from analyzing purchase transactions, which eventually must be captured in these systems to maintain inventory and financial integrity.

These same systems can minimally provide rudimentary quality measurements in terms of rejected parts or materials returned to the vendor. However, vendor performance management that is typically supplied by ERP systems or custom reporting has limitations. Typically, it is generated on a periodic basis, which seldom is frequent enough. Data reported tends to be averages computed over time and possibly most influenced by recent events. But it is difficult to detect trends early enough to counter them, and impossible to drill down from these summaries directly to the relevant data from which they are compiled. Where multiple back-office systems are deployed across a large multiorganizational enterprise, there is often inconsistency between operations that make it difficult to gain a consolidated view. Plus information is difficult to share both internally and externally.

Additional statistical data can be captured in specific quality management systems. These have traditionally been intense number-crunching applications that used to reside exclusively in the domain of the quality assurance professionals. However, with wider adoption of initiatives such as Lean Manufacturing and Six Sigma programs, there has also emerged a new generation of supplier relationship management applications. These systems are making their way, not only into the hands of the purchasing department, but also into the mainstream of quality and cost-conscious business management.

These systems are able to combine quantitative data from ERP and other back-office applications with external data from sources such as credit rating services, certification boards and financial reporting services. Some can also include qualitative data input from buyers and other personnel, recognizing that other factors that cannot be measured come into play in vendor selection and management, factors such as prior knowledge of vendors, or the ease in dealing with them. They not only can rate vendors, but they can substantiate these scores by drilling back into the transactional data from which they are derived. They make use of real-time data and analytical tools that are supplemented with sophisticated algorithms that allow them to anticipate trends. These types of predictive technologies are described in more detail in Chapter 7. And finally, this new generation of applications makes use of visualization tools that translate raw data into a format that is immediately understood, not only by quality assurance statistical gurus, but also by typical business managers.

In extending your vendor management efforts along these lines, look first and foremost for the ability to:

- Capture and analyze data in real time
- Drill down to specific data from summaries that catch your attention
- Consolidate data from multiple internal and external sources
- Apply analytical and predictive tools to anticipate trends proactively

THE DIFFERENCE BETWEEN DIRECT AND INDIRECT MATERIALS

The transactional buying aspect of direct materials has been largely automated in companies that have fully implemented both MRP and EDI. However, while many companies today have purchased MRP systems, full deployment of these systems is usually overestimated. In addition, except for large companies, very few have fully implemented EDI with all their direct material suppliers. Nonetheless, tools for the automation of this process have existed for many years, while the purchase of most indirect goods remains very much a manual process.

There are in fact many differences between direct and indirect materials. While direct materials are used in production, and therefore can be planned for with MRP, except for some maintenance consumables, planning for indirect goods is generally done manually, if at all. The accounting for the two is different. The cost of direct materials contributes to the cost of goods sold (CGS) while indirect materials are charged against general and administrative (G&A) costs. The logistical requirements are different. Product delivery delays of direct materials can cause production delays, interruptions or stoppage. Again, except possibly for some maintenance and repair parts, delays result in less of an impact. Typically, companies exercise a large degree of control over the purchase of direct goods, but are more at risk with minimal controls exerted over indirect materials. Yet conversely, the processing costs of purchasing indirect materials far exceeds that of direct.

But it is primarily this difference in the processing of orders that causes us to look separately at the automation of direct material purchase and that of indirect goods. In evaluating applications for use in indirect purchases you must consider their ability to manage a catalog, manage the process with appropriate controls and workflow and provide adequate reporting and analysis. In addition you need to consider how well they scale as you may be adding a significant number of users to your online systems. Yet in evaluating applications for the purchase of direct materials, first and foremost you must consider how you will integrate it with your existing MRP planning on one end, and your back-end purchasing and financial systems on the other. Add in the possible integration with strategic sourcing applications, and you see that integration is truly the prime consideration.

TRANSACTION AUTOMATION OF DIRECT MATERIALS

Those companies that have successfully implemented EDI are already conducting E-commerce. For them, conducting Internet-based procurement will require more of a migration strategy than a new E-commerce initiative. They have a significant head start since EDI has already been integrated with their back-office applications. Introducing Internet-based transactions can significantly increase the percentage of their total supplier base that can benefit from the improved speed and efficiency gained by electronically enabling these transactions. Not only does Internet-based trading further reduce the costs and latency by eliminating the value-added-network (VAN) component, but it paves the way for smaller trading partners that could not previously afford the VAN-based services. (See the section labeled "B2B vs. B2C" in Chapter 4 for an introduction to EDI and VANs).

If a company today has implemented EDI with 20% of its vendors, EDI-based transactions may account for as much as 80% of the company's spending, but it is certainly not automating 80% of the documents exchanged. Therefore, even large companies that have made a serious investment in EDI can see large gains from leveraging the Internet in these transactions. And those companies that had previously seen EDI costs as prohibitive can now enter the domain.

The first phase of migration or initial implementation of Internet-based procurement of direct materials is the machine-to-machine communication of orders. Yet the real benefits are realized only when this process is directly integrated with your ERP or other back-office systems. Almost every commercially available MRP system today automates the conversion of the demand generated by MRP to purchase orders to a primary vendor. Automating the process of transmittal of those orders to the prescribed vendors is obviously the first step in integrating E-procurement with ERP. The degree of automation that can be achieved throughout the next steps in the process, including order confirmations, receipts, shipping, invoicing and payment remittance will be determined both by the technology of the solutions implemented and by the type of relationship you have with your suppliers.

The desire to further streamline and automate these processes is at the heart of many supply-based reduction efforts. If you recall from Chapter 4, Dell's criteria for supplier rationalization were based on fulfillment capabilities, that is, the suppliers' ability to respond within minutes of identified demand. With relationships this close with a small number of trusted suppliers, formal order confirmation is nonexistent and receipts can be delivered directly to the stockroom or the production line with little or no matching required for payment processing. The elimination of these steps, in of itself, serves to streamline the entire process. No transactions can be eliminated from the back-office systems. Receipt transactions must still be created to record the inventory asset on your books. Invoices must still be created to record the liability to your vendor. And payment must still be processed to reduce your liability and increase your cash assets. Yet any or all of these transactions may in fact be initiated from one physical event. The formality of receiving the item may cause the inventory to be both incremented and immediately issued to production. It may also automatically generate payment, which in turn creates and clears an assumed invoice. Or an invoice received electronically may automatically receive the inventory and issue it to the production line. For all but the largest of companies or the most tightly knit business communities, this type of automation is extremely rare. However, this example does serve to point out that the number of different touch points between E-procurement transaction-based applications and your ERP system can be high, and it can vary significantly depending on the level of full E-business integration you have achieved.

There have always been trade-offs between selecting best-of-breed point solutions, or a better integrated, but possibly less robust solution from your ERP vendor. In the case of procurement of direct materials, the advantages are clearly weighted in favor of a more integrated solution. Even if the solution available from your ERP vendor does not completely satisfy your needs for indirect procurement, it is worth careful consideration, since it is unlikely that the same E-commerce application will serve both your direct and indirect procurement needs.

TRANSACTION AUTOMATION OF INDIRECT MATERIALS

While integration with back office systems was the critical issue in the E-procurement of direct materials, catalog management, workflow automation and control mechanisms are of prime concern with regard to indirect materials.

By now, most everyone is familiar with the experience of browsing through online catalogs. Two- and three-dimensional images and intuitive user interfaces are the first considerations that come to mind. Yet these are more important considerations to a B2C environment than they are for B2B. Much more important considerations to implementing a B2B E-procurement application for indirect purchases is not so much the pretty pictures the users see, but how the content of those catalogs are managed.

One of the lures of indirect E-procurement is making the purchase of materials from approved suppliers easier than bypassing controls put in place to control spending. This means putting an electronic version of the vendor's catalog on the desktop of the user. The content of that catalog includes product codes, descriptions and pricing, as the bare minimum, but may also extend to specifications, visual images and other multimedia content. How often does this content change? The answer to that question will vary enormously dpending upon the items purchased. Number 2 pencils change far less frequently than laptop computers.

Catalog management used to mean making sure you threw the old catalog out when the new one arrived, in order to assure you had the latest and most accurate information. It meant filing it in an appropriate place which made it easy to find when needed. Yet, no matter how carefully this catalog management was performed, typically the purchasing agent checked with the vendor, usually by phone, to make sure the catalog item was still valid and available, and that the price shown was accurate.

One of the goals of E-procurement is to remove that burden from the purchasing agents, and the most accurate and up-to-date information is immediately available to these buyers on their desktops. But whether it is done manually or electronically, someone still has to do it. This is the hardest element of Internet procurement to manage and is a significant contributing factor when these initiatives are not as wildly successful as expected.

Obviously you will be managing content from multiple sources, even if it is presented to the user as a single catalog. Since your vendors are your source for this information, having them manage the content is your best bet and providing them with flexibility in this process is the key to making it happen.

If your supplier sends you updated content in order that you maintain a copy of it behind your own firewall, the greater the degree of flexibility of data formats you can accept, the more suppliers can be incorporated into your initiatives. However, the alternative approach is to supply your buyers the ability to "punch out" to the vendors' catalogs, managed and maintained on your suppliers' servers. While this requires additional feature functionality and technology capabilities from your E-procurement application, it has an enormous impact on the level of effort necessary to manage your E-procurement implementation on an ongoing basis.

The second important consideration for E-procurement of indirect materials is workflow automation, and the complementary ability to define rules and monitor and trigger events. It is the combination of these capabilities that provides you the level of control you need to reduce maverick spending and streamline the approvals and purchasing process. Unlike the procurement of direct materials, for which the generation of demands and the selection of vendors are automated by MRP, the manual processes of acquiring the indirect materials require an additional level of control.

Obviously the job of workflow automation is to direct tasks to the appropriate individual and monitor the status of those tasks. The more the process can be directed through the application of business rules, the more value it can bring. Combine the application of business rules with the ability to "listen" for events, generate alerts and redirect process flows and you begin to see genuine results.

Once the process is fully under control, purchasing cards are a means of further streamlining it by closing the loop through to the payment process, particularly for low value transactions. According to the National Association of Purchasing Card Professionals, the transaction cost of making small dollar payments is disproportionate to the value of the purchase. The cost of processing a $25 payment is the same as the cost of processing a $100,000 payment. As a result, the cost of payment can easily exceed the value of the item purchased.

Purchasing cards can simplify the process and drastically reduce the cost. Both the supplier and the buyer benefit from this. The buyer reduces the number of checks written and diminishes the effort associated with collections. That is, in fact, what you pay the card issuer to do. The use of purchasing cards also changes the accounts payable (AP)/purchasing reconciliation process, simplifying it without sacrificing control over individual purchases. The supplier benefits by getting paid faster and by eliminating the need to generate an invoice.

The purchasing card process, as depicted in Figure 5.6, is essentially a credit card adapted to a B2B environment in which transactions are documented and retained in the general ledger (GL). One note of caution, however, is germane. Not all suppliers are capable of passing the appropriate level of detail to the credit card company, possibly limiting your ability to resolve discrepancies easily. Therefore it pays to carefully consider which suppliers you will include. However, the more universal the application of this payment method, the more the entire purchase cycle is streamlined.

Scaleability is also an added consideration in managing Internet procurement of indirect materials. As previously depicted in Figures 5.4 and 5.5, the number of users directly connected to the procurement cycle can scale up dramatically. So must your E-procurement systems.

SUMMARY

E-commerce is just a step in the journey to full E-business integration. But not only is it a giant step, it is a multifaceted one. It requires careful

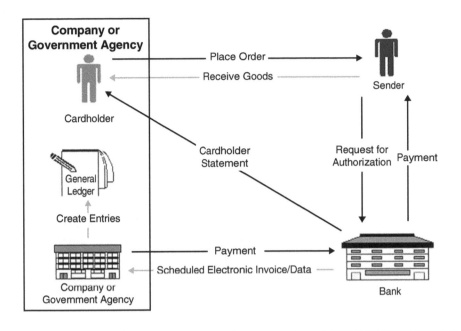

Figure 5.6 The procurement card process. (Source: National Association of Purchasing Card Professionals.)

consideration of current business relationships, internal technical assessment and strategic priority setting. Your sell-side E-commerce strategy must be tightly linked to your overall channel strategy. Your business model, whether direct or intermediated, must be determined based on product mix, industry trends and technology. You may need to carefully examine the requirements and benefits of trading exchanges, including those which are independent, collaborative or private.

The buy side of E-commerce can offer exciting opportunities to reduce processing and purchase costs, along with procurement cycle times. It can free up your buyers for more strategic activities, such as vendor selection and relationship management. Yet if you underestimate the integration efforts to connect to your back office or the effort of catalog management, your results can be disappointing.

Although just a step in the longer journey, E-commerce is one that is crucial for survival in the emerging E-business world.

6

INTEGRATING ALL INFORMATION ASSETS

As we continue to explore the extension of ERP for the express purpose of maximizing profits in the world of E-business, we seem to come back again and again to the topic of integration. More and more companies today face a significant challenge in integrating multiple business applications. This is the natural result of fewer and fewer companies running a single, all-encompassing business application, either in a lone facility or across multiple sites of a multinational, multiorganizational enterprise. The inability to integrate leaves an incomplete and/or disjointed view of your enterprise.

In addition we have seen the growing need to interoperate effectively within an integrated business community. The inability to integrate can prevent you from actively participating with the most successful value chains. In our definition of full E-business integration we uncovered the need for new business models that involve multiple companies working cooperatively and collaboratively together, in a seemingly seamless manner, as if they were a single, *virtually* vertical enterprise. All these factors only increase the importance of integrating all your information assets. And successful integration paves the way for profits.

This chapter examines this growing need for enterprise application integration, and as we do so, we uncover the magic number three. There seem to be three general reasons why this becomes a critical issue, three levels of integration that need to be considered, and three alternative approaches to the problem.

WHY IS INTEGRATION AN ISSUE? REASON 1

For many companies today, the collection of business applications that form the foundation, upon which they run their business, has grown over time. Few of these applications date back to what I would call the BERP era — no pun intended. BERP stands for Before ERP. Some of us in the aging population today can recall back to the days when packaged software of any kind was viewed with disdain for any but the smallest of companies. That was back when the IT department was known as "data processing" and software developers, architects and engineers were simply known as programmers. Back then when companies of any substantial size at all needed an inventory system, or an accounts payable system, they either hired programmers that designed and coded the application or they hired companies that would come in and design and program it for them. I essentially just described the first two companies I worked for right out of college, and thereby probably dated myself even more than my gray hair does.

And the reason most of those applications don't exist today is because most of them never made it past Y2K. Years that didn't begin with "19" weren't something you worried too much about in 1975. Yet while few applications date back to the BERP days, there are many applications that may be known as "legacy" systems that operate at the very core of enterprises — systems of a certain age and maturity that may not take advantage of the latest technology available. They may have been home-grown or they may have been purchased. They may be individually developed applications that needed to be interfaced or integrated with other applications, or they may be early versions of ERP systems. They may be rudimentary or they may have grown to be comprehensive, mature and feature rich over time.

So this is the first reason for the growing need for enterprise application integration. These legacy systems — whether they were developed as individual applications or as an early, integrated ERP system — simply did not have a broad footprint. They have never done everything that integrated systems can do today. This has led to a proliferation of applications over time. If the engineering department's needs were not met, it looked elsewhere. Or the sales or service departments. Or human resources or any number of other departments within a single company. Whether this proliferation was because of a conscious decision to purchase, install and implement "best of breed" (BOB) systems, or because there were maverick departments that went their separate ways, the result was the same.

This problem is further aggravated by the fact that business requirements have become more sophisticated over time, creating the need for extensions to the typical ERP solution. These might be applications for Supply Chain Planning (SCP) or Supply Chain Management (SCM), Customer Relationship Management (CRM) or Product Life-cycle Management (PLM). Those companies that implemented ERP systems did so with the expectation that the ERP system would continue to grow with them, expanding to satisfy new requirements. And in fact most ERP vendors today are striving to extend their solutions with new products or new modules. But the reality of the situation is that sometimes business needs grow more rapidly than the ERP vendor is able to respond, since these vendors are constrained in two ways. First of all, they are constrained by the technical architecture of their existing products. While development tools and the technology that drives business applications have matured significantly over the past years, new tools are usually incompatible with existing systems and architectures. Second, ERP vendors are constrained by their installed base of customers. On the surface this may seem like a contradiction, but in fact users of enterprise applications typically demand their existing applications be enhanced with user requested features to existing modules and products.

This enhancement process has the potential over time of becoming more and more difficult. As the ERP vendor's system grows in scope, it becomes more difficult to touch one area of the application without impacting other areas. Therefore, directing the enhancement of the entire system becomes akin to steering a battleship.

It is exactly through this set of circumstances that point solution applications have emerged. If the ERP system is a battleship, point solutions are the PT boats, quick to be launched and easily maneuvered. As a new business requirement is recognized, point solution vendors can respond much more quickly than an existing ERP vendor. First, they have the luxury of choosing their development environment and technical architecture, taking advantage of the latest and greatest in Web services and other technologies. They are not constrained by legacy architecture. Nor do they have existing customers to satisfy with bug fixes and enhancements to installed software.

Point solution vendors enjoy a first to market advantage. Depending on their goals, target market and early level of success, they may or may not continue to develop their solutions as BOB applications. Many will start as small companies with a bright idea and a viable plan, but they may not achieve a magnitude of success quickly enough to ensure continued self-funding or outside capital investment to take them to the next level of a Siebel, for example. Siebel is perhaps the most widely recognized BOB vendor in the world. It has become the world's leading vendor of customer-facing software and the BOB for CRM (Customer Relationship Management).

From the emergence of companies like Siebel and other BOB vendors has arisen the age-old dilemma of choosing between BOB and extensions from your incumbent ERP vendor. First, there is the question of timing. Will your ERP vendor have a module, feature or new product before your need becomes critical? Then there is the trade-off between the additional functionality typically available from a BOB software supplier and the level of effort required to interface or integrate the solution with your other applications.

Such is the case for BOB vendors like Siebel, which have extensive functionality and immediate availability, but usually also come with an expensive price tag. But for every Siebel, there are dozens of other small companies that create a product, are perhaps among the first to market, yet gain only a modest market share. These alternatives are typically more attractively priced, although probably not as fully functional as a BOB. In their quest for additional market penetration they typically seek partnerships with ERP and other software vendors with established salesforces. The advantage of these partnerships to these point solution vendors is obvious. But what is the advantage to the ERP vendor? The lure for the established software vendor is the ability to extend its product solution almost immediately, by-passing the majority of the time and effort of extensive research and development projects.

In purchasing one of these complementary products from your ERP vendor's partner, you would logically expect the issue of integration to be ameliorated. Unfortunately, in the software world these partnerships often progress no farther than the joint press release. Even when two vendors offer an integrated solution, this integration falls far short of customer expectations.

Many times initial "quick and dirty" interfaces come with the promise of more extensive and seamless integration in the future. More often than not, however, these promises of seamless integration go unfulfilled. Why? One simple reason is because these integration efforts between two software companies are never as simple as they appear to be at first glance. Another contributing factor can be because after the first blush of success fades, with several new software licenses sold, the partnership simply does not yield the kind of revenue results expected for either party and the enthusiasm wanes. The other possibility, of course, is that the ERP vendor simply used the point solution as a stopgap measure until it was able to provide a fully integrated solution of its own, one that involves no partnership agreements or royalty payments.

In order not to portray the ERP vendor as the "bad guy" in this situation, it is necessary to point out that it is equally likely that the point solution vendor will continue to enter into other partnership agreements. One of these could potentially lead to a situation of conflict of interest, or more likely, acquisition by a competitor of the ERP vendor. This leaves what was initially a fragile relationship in a potentially dysfunctional state.

So while there are many paths leading here, the destination is the same, the proliferation of individual applications and the potentially great need for integration.

WHY IS INTEGRATION AN ISSUE? REASON 2

The second reason for this growing challenge has resulted from the merger and acquisition frenzy of the past decade. Many companies have grown predominantly by acquisition. And with the acquisition of a company comes all the software that was installed and implemented at that company. In some instances these companies were fully integrated into the acquiring company, in which case migrations and reimplementations were the name of the game. This was the case for Belden Brick. Belden, as you will recall from Chapter 4, is a quality leader in architectural brick. In an effort to strengthen its position in a consolidating market, Belden acquired Redland Brick to gain added economies of scale and maintain its leadership position. Belden valued Redland's brand identity in its separate markets, yet needed to leverage information within both companies to take full advantage of the synergism between the two. At the time of the acquisition, Redland Brick had just completed a successful implementation of an ERP system, but Belden Brick was still struggling with a partially complete, and not entirely successful implementation of a different ERP system. In this case the ERP strategy for the newly merged enterprises proved to be the replacement of the partial implementation of the acquiring company's ERP system with the same ERP system its acquisition had successfully up and running. Which ERP system was kept and which was discarded proved to be far less significant than the approach. The fact that Belden Brick recognized the need for executive commitment and objectivity in the selection process is a tribute to its management.

This was exactly opposite to the case of the anonymous valve manufacturer we met in Chapter 3, which will remain nameless once again. If you will recall, this company embarked on an ERP implementation without any kind of commitment to change, and therefore it unnaturally and unnecessarily constrained its new system to behave like its old, ineffective system. Before the company completed the implementation of this new system, it was faced with a new challenge. Its parent company decided to consolidate business units and therefore it was merged with another division. As it was the larger of the two operating units, this company was favored with the status of being the acquiring division. The business unit that was acquired had successfully implemented an early version of an ERP system, and had been productively running it for many years. Yet, in contrast to the objectivity demonstrated by the management of Belden Brick, the decision here was to discard the successful implementation of the inherited system for the new, unproved and unsuccessful implementation of the acquiring division.

I wish I could say that my experience has proved that this kind of decision is typically reached with the objectivity and due diligence exhibited by Belden Brick. However, more often than not, in a contest between "your ERP or mine?" the acquiring enterprise is typically the winner, for better or for worse.

But mergers and acquisitions do not always result in this kind of operational integration. In many instances these acquisitions remain as separate and autonomous operating units, in which case the inherited systems may survive. And now corporate management has a problem in gaining a unified view of all its operating units and interoperability between divisions is a challenge, to say the least.

This is the situation in which we discovered Myers Industries when we first met CEO Steve Myers and CIO Andrew Winer in Chapter 2. If you recall, Myers Industries has grown by acquisition. Today it operates 25 manufacturing facilities in North America and Europe and 42 distribution locations in over 31 states, resulting in the operation of many different ERP and other related business information systems.

Because each operating unit or division of Myers is allowed a degree of autonomy, both in plant operations and information management, the result was a proliferation of business systems, which was preventing Steve Myers from gaining a consolidated and unified view of his business. It was also creating a road block to fully leveraging the business strategy of acquiring companies with complementary products. Without that consolidated view of customers it was impossible to exploit all sales opportunities across the combined divisions' base of customers.

WHY IS INTEGRATION AN ISSUE? REASON 3

Finally, the third reason behind this challenge is the emergence of E-business, and the changing paradigms that this emergence brings. We explored this phenomenon in detail in Chapter 4, where we saw the evolutionary process bring us through the generations of E-information, E-commerce and finally full E-business integration. Among the conclusions drawn was the observation that no company today can operate in isolation or even at arm's length from customers, suppliers and partners. Boundaries between businesses are blurred. Successful E-businesses of the future will be those that treat E-business as the collection of processes, which allow multiple companies to work cooperatively and collaboratively to produce a seemingly seamless integration of businesses operating as a virtually vertical enterprise. And with this integration of business processes comes the requirement to integrate disparate business applications.

So, whether the need for integration arises from the proliferation of business applications within your own enterprise, from the results of mergers and acquisitions or from the demands of E-business, integration emerges as a significant challenge in responding to the demands of business today. What then constitutes integration and how do you go about meeting these challenges?

WHAT CONSTITUTES INTEGRATION?

Oddly enough, just as there are three predominant reasons integration is an issue, there are three aspects of integration. The most basic requirement of integration is at the data level. The second aspect is perhaps the most complicated and problematic level since it is the one that is buried the most deeply into the application. This second aspect involves the actual application logic itself. The third level is one of consistency and universality of access and visualization.

At the data level, there is master data, transactional data and data that might be viewed as somewhere in between. This "in-between" data is that which is typically stored with the master data but becomes dynamic as the result of transactions — inventory and account balances, allocations and reservations.

With the proliferation of applications comes proliferation of master data. Within a single company or division, having multiple copies of master data such as customer, supplier or inventory data, in of itself, is not the problem. The fact that the information resident in these multiple copies may be different is the problem. In a rather simplistic example, an order entry application, whether it is a back-office application or a Web-enabled electronic storefront, must have access to customer and credit information and inventory availability. Supporting master data can potentially reside in an order entry system, an inventory system and an accounts receivable system. Theoretically ERP took care of this problem with the introduction of a single integrated database. Theoretically yes, but universally? No. The comprehensiveness of ERP implementations in general has been overestimated. Many companies that have purchased ERP systems are still running a hybrid mix and as the push to Web-enable legacy systems increases, this condition will grow. The need to share common data between applications will grow along with it.

In a multiorganizational environment, particularly one that has grown and developed either by acquisition or with a planned level of autonomy, the goal is typically to gain access or a view of data across multiple sites and applications. In our simplistic example, the inventory availability may be from any number of warehouses or manufacturing facilities, across multiple operating units or companies.

But then there is further integration that goes beyond the data level and must involve application logic. For example, our centralized order entry function, which has access to multiple sources for inventory, may need to apply certain logic to determine the priority sequence of locations from which to pull inventory. Do you always ship from the closest warehouse? Do you always ship from a certain location unless there is no availability? If so, how do you select the alternative location? These examples require more than simple sharing of data. They require logic that applies business rules, decisions and policies.

Finally, there is the level of integration that resides in the presentation or visualization layer. When integrating multiple disparate business applications, do you concern yourself with a common and consistent look and feel? And how is access achieved? When there is a combination of legacy system with new Web-enabled applications, do you front-end the legacy systems with a browser-based front end? There are several products on the market today that provide you with the ability to bring legacy applications to the Web. This is generally the least disruptive approach in that the underlying application remains the same, along with the application logic and process flow. This may be an acceptable alternative if in fact the existing "green screen" application generally supports your business processes. But it does nothing to improve the underlying application, so if the overall objective is also to provide additional functionality, you may be better off supplementing your existing application with a new Web-enabled application that extends the feature/functionality of your existing system. Just how far you go will be determined by what you are really trying to accomplish.

WHAT APPROACH DO YOU TAKE?

The temptation, when dealing with multiple disparate legacy systems, will be to rip out these existing systems and replace them with a single, standard, comprehensive state-of-the-art application. This would effectively circumvent the problem by eliminating the need for expensive integration efforts. In the case where the need for integration is the result of acquisitions, this in fact may be the best route. This seemed to be the general direction many larger, multidivisional, multicompany enterprises were headed toward at the end of the last millennium. The feeling was, "once we get past Y2K, we'll see about replacing all our old systems." This included many companies that had made an investment in SAP's products at a corporate level and were interested in recouping some of the spent cost by leveraging that investment across multiple divisions and operating units that previously had made autonomous decisions concerning the implementation of systems. Oracle supported this strategy by running advertisements that said, "You will never become an E-business by piecing together software you already have." The alternative, Oracle implied, was to implement its extended ERP solution throughout your enterprise.

So, yes, this is one approach to dealing with the growing challenge of integrating a proliferating set of disparate business applications. It also happens to be the one that tends to be the most expensive and the most disruptive to your business. Many companies, like Myers Industries, once past Y2K, began serious investigation of this approach and got instead a serious dose of reality. Anyone who has experienced, firsthand, an ERP implementation knows that the software cost, even when it is coupled with the cost of external implementation consultants, is but a fraction of the true cost in terms of the blood, sweat and tears expended during the course of the project. And that doesn't even include the cost of any necessary tailoring or outright customizations to the software.

This dose of reality also caused them to look more critically and carefully at what may have been aging legacy systems. Hence, the term *heritage* system was introduced. Heritage systems can be defined as legacy systems you are proud of. And if you are proud of that system, then chances are it is doing the job that it was intended to do for you. Many companies realized that an enterprise-wide effort to rip out all existing legacy systems, to replace them with newer ones, would cost them hundreds of thousands, if not millions of dollars. Ignoring for the moment all the blood, sweat and tears, the net result of this expenditure would have been simply to get back to exactly where they are today.

So let's assume you decide, instead, that the existing legacy systems will stay put. This decision of course makes the assumption that they are in fact meeting the majority of your business needs, which is a critical factor in the equation. Let's assume for the moment that these systems meet a large portion of your basic business needs, but as you continue to transform your business into an E-business, as is inevitable for all organizations today, your needs will grow and change. So not only are you still faced with the requirement to integrate business applications, this need will escalate over time as you seek to supplement the functionality of your existing systems. Now what do you do? What alternative approach could you take to a mass replacement of your applications?

The most obvious answer might seem to be to write custom interfaces between your existing systems, and in fact this was indeed the most common approach of the 1980s and 1990s. Unfortunately this too is an expensive proposition, because this type of integration is not a one-shot deal. All but the oldest homegrown legacy systems continue to grow and change over time, particularly commercially developed ERP systems. ERP vendors must continuously expand their solution footprint in order to remain viable, and changing business conditions continue to present new problems to solve and functionality to be provided. As your underlying systems grow and mature, the interface requirements are subject to these factors, causing you either to delay in implementing the newest features your maintenance dollars are buying you or to spend additional time and money to analyze, design, code and test the interfaces.

The third, and least disruptive alternative is to consider "technology enabling" your existing applications and using this process to achieve integration. Sounds pretty good, but what does that really mean? To answer that question, we have to return to the different levels of integration — data, application logic, and presentation or visualization.

Using this alternative, how do you assure that the information is consistent between complementary applications? The ultimate goal should be to have a single copy of any particular piece of data. How achievable this goal is will be dependent on a variety of factors. Let's say you are a single company with an integrated ERP system installed, but you have recently implemented a new Web-based order entry system to enable distributor and possibly customer self-service. Chances are this order entry solution came with its own customer master file. But, of course, you have a customer master file maintained within your ERP system. The first consideration should be to choose this new E-commerce application carefully. It was never meant to be a stand-alone application operating in a vacuum. How easily does it interoperate with back-office applications?

For example, Oracle's sell-side E-commerce application has been designed to interoperate seamlessly with its own Oracle ERP application. interBiz, formerly the E-business applications division of Computer Associates and today a product group of SSA Global Technologies, has taken a different approach. interBiz Store, its Web-enabled E-commerce application, defines the data elements it needs in a common object model. interBiz then used wrapper technology, described in Chapter 2, to expose those data elements from its multiple ERP applications to that object model. By referencing the data through the common object model, it is not replicating data, but simply redirecting interBiz Store to access it directly from the back-office application, eliminating the need for redundant data and synchronization. interBiz Store is used as a means to technology-enable the back-office ERP applications and in doing so reduce the need for specific interfaces. Companies with applications other than interBiz ERP systems would simply need to develop their own wrappers to expose this same data to interBiz Store.

interBiz Store, and other, similar E-commerce applications, can also be viewed as a presentation or visualization layer to existing legacy systems.

Once installed and implemented, the underlying applications can be reengineered or replaced more transparently to the users.

A similar approach can be used in integrating disparate business applications, all of which may be legacy systems that may have proliferated at any particular company. An apparent solution to providing a single copy of any piece of data would be to have all data physically reside in a common repository, but this can have significant impact on any or all of the disparate applications. Data must be migrated, duplications resolved, and applications must be modified to redirect access and recognize new formats and structures. By using the approach of a common object model and wrapper tools, this effort does not disappear. However, it is not only significantly reduced, it paves the way for future changes by not locking any business function into any specific application, and by not locking any application into any specific data format or structure.

This may seem fairly straightforward when it comes to the data and presentation levels of integration, but how can you "technology-enable" applications to address integration issues where application logic is required? The answer is actually simpler than you might imagine. You remove the logic from the application. This is a very important concept, and one that was introduced earlier in Chapter 4 in the discussion of AMR Research's Enterprise Commerce Management, in the context of the layer it labeled as interaction services. However, due to its significance here and the fact that some of our less technical readers may have found this explanation more than they desired to know, it bears repeating.

Traditionally, business applications have embedded business rules and decision making within the logic of the application. By removing these rules from the applications themselves and managing them in a separate rules engine, this paves the way for greater flexibility and creates a more agile environment. When the rules change, the policy criteria, such as thresholds and limits, can be changed without having to touch the application itself. Clearly, it is much more difficult to retrofit this type of functionality into existing applications, as it means gutting the application itself. It means replacing internal program code with external calls that allow for a much more flexible range of results. However, most businesses today have not implemented their last application. Particularly with the demands for outward-facing applications required to interoperate within and throughout the full supply chain, there is plenty of opportunity for companies to take advantage of this type of technology. And, let's not forget what brought us to this discussion in the first place — integration. In defining how two or more applications interoperate, there is a great deal of potential for utilizing a combination of technologies, which are essentially external to the applications. And they may only need to be directly connected to the applications at select, critical points, or perhaps not at all.

Workflow engines have provided excellent vehicles to begin to address this goal of removing logic from the application. The entry of a new customer can alert the credit manager to initiate the credit approval process, thereby expediting the processing of a quotation and the generation of a booked order. The receipt of the order can automatically release purchase and production orders. The receipt of a service call can notify the assigned engineer. The transmittal of a loan application can alert a loan officer. The consumption of stock can initiate a replenishment order. An account balance running over budget can alert the responsible department manager.

As you can see, on the surface, some of these examples may have nothing to do with integration. If you manage customer credit using the order management and the accounts receivable modules of your ERP system, these should already be integrated. If your ERP system is supported by workflow technology, then the process of taking an order, and other activities, should trigger the credit check. What happens, though, when, as in the data level example above, you determine it is time to launch a B2B E-commerce initiative and you decide to implement a sell-side E-commerce application to accept orders over the Internet? Now you not only need to feed orders from your Web-based order entry application to your back-office order management system, but you also need to trigger that credit check that combines data from all three modules. How can you trigger that check to occur without writing a specific interface between the two applications?

The answer to this question lies in the ability to perform some level of "event management." While potentially a very sophisticated and powerful technology, I will present it in conceptually correct, yet grossly simplified terms. The first aspect of event management that is required in this example is that of "event listening." In this case, an event listener, which is not necessarily a function of either the new Web-based order entry application or your back-office ERP system, simply waits for an event to occur. In this case the event is the creation of a new order. Once this event is detected, another event must be triggered. This other event may be as simple as the creation of a task in a credit manager's "to-do" list, in which case its job is done. The order will be taken, but essentially held until credit is approved at some later time. In other cases, the event that is triggered is to automatically initiate an online credit card approval process performed by a third-party software provider such as Paylinx or CyberSource, credit card service providers that actually obtain credit card authorization from the bank or other source of credit.

As you can see, by using workflow technology to combine technology that applies business rules and business policies together with an underlying transaction-based application, you can start to orchestrate business processes that may span multiple applications.

Let's consider another example. You are a manufacturer, which has grown by acquisition, with three different manufacturing facilities. The three facilities for the most part produce different product lines, but there are some common products among the three. There is some overlap between the customer bases of these three, because each of the plants addresses somewhat overlapping product lines. There are three different ERP implementations with three different customer master files. To reduce cost and streamline operations, you have consolidated the salesforce and you are implementing a single Web-based order entry system. You enter a new order from that system, referencing the customer using the customer number one of the back-office systems knows that customer by. Using a common object model that has been defined and a wrapper that exposes data in the ERP system, you check product availability and find that you do not have inventory enough in stock to ship. Behind the scenes, rules are defined that do some mapping and translating of customer numbers across the three different ERP systems. In this way you determine if in fact this is a customer known to either of the other two systems and it uses workflow to initiate a check on inventory availability in both locations. It finds it has sufficient inventory in either of the locations, but that customer only exists in one of them. So predefined business rules determine the selection of the warehouse from which to ship. The order is accepted. Workflow technology then is used to create an order in the back-office sales order entry module of that ERP system and the transactional flows associated with that system take over.

By complementing existing business applications with advanced technology such as rules engines and workflow, you again make it easier to reengineer or replace the underlying applications selectively as necessary to continue to meet the challenges of the ever-changing business world.

SUMMARY

The requirement to integrate diverse information assets continues to grow, whether it stems from the proliferation of business applications, growth by acquisition or the need to interoperate within an integrated business community. Integration can be at the data level or require the application of business rules, policies or basic logic.

Whether the desired result is shared data, shared logic or simply a unified or consolidated view of your business, you will be faced with choices. You may wind up replacing systems or writing custom interfaces. Or you may turn to enabling technologies to blend old technology with new in order to achieve that all-important, emerging metric of interoperability.

7

PREDICTING BUSINESS EVENTS

As the pace of business and the rate of change itself accelerate, there comes a point in the life of your business where efforts to streamline activities, reduce cost and respond more quickly meet the law of diminishing returns. It will become more and more difficult to justify the cost of these efforts for what may be an insignificant return on investment. Theoretically, at least, you will be operating at maximum speed. But what if your speed and agility at this point prove simply to be not fast enough to achieve or sustain market leadership? What next?

It is becoming more and more critical for business executives to focus on the future and predict business events. Conventional performance metrics have focused on the past, and at best might give an occasional fleeting glimpse at present status. Pioneering executives today turn to predictive technologies, commonly referred to as artificial intelligence or neural learning agents, to complement and supplement their analysis capabilities. They employ these tools to anticipate scenarios and practice "what if" strategic management. This chapter explains this technology in lay terms and suggests how it might be applied in a business context, not in the year 2020, but here and now.

After reading the first six chapters of this book, you will undoubtedly have learned a little bit about me, the author. You will have learned a bit about my background and expertise. You will have discovered some of the companies I have worked for and you will have even been introduced to a couple of my friends. Yet before I delve into the last of the technologies I feel you need to understand to compete in today's mixture of business and E-business, I would like to take a moment to introduce you to another facet of me and my life, so as to better explain the importance of this technology and the value it can bring.

I admit to leading a double life. By day, I am a mild-mannered businesswoman. You will see me trudging through airports among the other worn and weary business travelers. You will see me in corporate offices or manufacturing plants. You can hear me speaking at conferences about applying software solutions to the business of manufacturing. You will recognize me, dressed in a dark, conservative business suit and low-heeled shoes, with graying hair and laugh lines around my eyes. My credentials can be displayed on walls of my office — a B.A. in math/physics, an M.A.S. in computer science, the usual awards and recognition by various Who's Who organizations.

But you may not recognize me as my alter-ego, for on many weekends and vacations my attire looks more like black pajamas than a business suit, and I wear my credentials around my waist, as well as on my wall. The other world I live in, besides the business world, is the world of martial arts. I was drawn to the arts relatively late in life for a martial artist. Many start at the tender age of 6 or 8, and are considered "seniors" by the ripe old age of 35. I began my martial arts career shortly before my 33rd birthday, and I have since achieved varying degrees of black belt rank in three different styles of martial arts.

For the past few years I have had the privilege of studying with a true master of the martial arts. As in the business world, the world of martial arts is replete with ranks and titles. While others of his stature in the arts call themselves "Grand Master" and the like, this man carries the simple and deceptive title of *Hanshi*, which is translated as "root master." He uses this title because he is dedicated to the preservation of the teachings of those who have come before him.

SPEED VS. TIMING

Hanshi is a large and powerful man who is capable of moving with amazing speed and grace for his size. Yet when you see him demonstrate his skills he seldom moves much faster than a notch above slow motion, regardless of the speed and power with which his opponent attacks. He does this in order to teach his students that speed is far less important than timing.

To drive this point home, Hanshi tells a story of two neighbors who have been invited to a wedding. The wedding is scheduled to begin at 11:00 A.M. The first man sets his alarm clock for 9:00 A.M. He has a leisurely breakfast, takes his time showering and dressing, and then drives safely to the church, never exceeding the speed limit. He arrives at the church 2 hours after arising, is seated and watches with pleasure as the bride floats down the aisle on her father's arm. The second man sets his alarm for 11:00 A.M. He bolts out of bed, skips breakfast, showers and dresses with lightning speed and races across town, running traffic lights and breaking every speed limit he encounters. As a result, he enters the church at 11:30 A.M., to hear the minister pronounce the couple husband and wife and find he has missed the entire ceremony.

The second man moved four times faster than the first, but he still missed the wedding — because he started too late.

Applying this story to defense against an attacker, Hanshi teaches his students to anticipate what their attacker's next move will be. He explains that if you react only to where your attacker is at the start of the attack, you must move more quickly than the opponent in order to block or strike effectively. However, if you anticipate where the opponent will be at the completion of the move, you have much more time to prepare a defense or counterattack.

So, how can businesses today plan an effective defense? Who and what confront them as attackers? Anything that can potentially prevent a company from being successful must be viewed as an opponent, whether that is its competition, market conditions, political and economic elements or simply a time factor.

The most pervasive opponent operating against business today is time, which makes Hanshi's analogy all the more applicable. The pace of business today is accelerating at a frightening rate. The faster companies move, the more their customers demand more customized products within shorter lead times. As companies reach their maximum speed limit, timing will become more and more crucial. However, orders seldom come in the mail like wedding invitations, allowing many weeks of planning and preparation.

Whether you directly supply a consumer market or fulfill a role within the supply chain, this is not a new problem for companies. Customer demand within manufacturing lead time has forced companies to manufacture to forecast demand, and what is a forecast except a prediction of the future? But, of course, everyone knows that the one thing you know for sure about a forecast is that it will be wrong. It is the degree of "wrongness" that is the variable.

CLUES TO THE FUTURE

For beginner martial artists, their ability to anticipate their opponents' next move is generally a function of how much the attacker telegraphs that next move. Opponents will look directly at the target, or draw their striking arm back as if winding up for a punch. Certain obvious anticipatory moves signal a repeated pattern of movement. But when you watch a master like Hanshi, it appears that he knows the attacker's next move even before the attacker does. How is this possible?

The beginner martial artist perceives only what is both visible and obvious. Hanshi sees the less visible subtleties. When his opponent prepares to throw a punch, the arm is not the first part of the attacker's body that moves. If this doesn't make sense to you, just try throwing a powerful punch without moving any other part of your body. Depending on how the attacker is standing, the first movement may very well be in the toe or ankle. To the uninitiated, this may seem like movement that is entirely irrelevant to the strike. Yet the master knows that the first muscle that moves is the one that initiates the transfer of weight. While this may seem mysterious and confusing to most, it is the simple product of the application of the laws of physics to human anatomy. Of course, the movement may be so slight that it ordinarily goes unobserved. It is only through years of observation and training that it is detected.

If we transfer this thinking to our forecast, the fact that the forecast is always wrong may in fact be because it is based only on what is visible and obvious. Most forecasts today are based almost exclusively on history, perhaps with some factors for growth or seasonality thrown in. More sophisticated forecast methods may integrate data collected by demographics or customer profiling. But does the forecast take into account other external factors like stock market performance, catastrophic disasters such as earthquakes and floods, supplier performance, potential or current military conflicts, currency fluctuations or the latest box-office smash hit?

Whether you are predicting demand for product, general market trends or the likelihood of a business event, when forecast accuracy is low, chances are there are one or more factors influencing the situation, which are not accounted for in the model. These factors themselves may not have been anticipated, or we simply may not have had any hard data on their impact. Yet we can all think of a situation where seemingly unrelated events can have a profound influence on future sales.

Shortly after the beginning of the Persian Gulf War in 1991, I visited a company that manufactured watertanks. Within days of Iraq's invasion of Kuwait in the previous August, the company could foresee the possibility of an increase in demand for its product, and worked up a contingency plan to ramp up production to work round the clock, 7 days a week for as long as necessary. Even before the troop build-up began in earnest the following December, orders started coming in and the company began to implement its plan. But what was a fairly obvious possibility for this company was somewhat less obvious for their tier 1 suppliers. What effect would a military conflict in a desert halfway around the world have on a coating or a fittings manufacturer? How early did these businesses "see" the connection and did they have a contingency plan that allowed them to add temporary capacity quickly? And how many businesses were caught totally off guard? What about businesses in seemingly unrelated industries? How far down the bill of material (BOM) and how many tiers of suppliers do we traverse before we encounter even short-term parts and materials shortages caused by this increase in demand? Certainly there will be a ripple affect throughout the supply chain and a corresponding ripple in other intersecting chains. What information did they use to predict accurately the amplitude and duration of the demand curve?

BEYOND THE CAPACITY OF HUMANS

Is it possible to anticipate all events and conditions that can affect demand either directly or indirectly, and to immediately assess both the likelihood of occurrence and the impact they might have on demand? It hardly seems humanly possible. But who says the detection and analysis has to be done by a human? Predictive technologies, or artificial intelligence, have been around for decades; yet traditionally they have been limited to certain specialized applications such as chemical processing and, more recently, computer network monitoring. But what prevents us from using them to predict business events and analyze the potential impact of these events?

Previously we were limited by the availability of data in some usable electronic form. But the Internet has certainly changed that. Today we suffer more from information overload than we do from the shortage of data. The power of predictive technology is in its ability to correlate and analyze massive volumes of data, discarding what it determines to be irrelevant. So while the human mind must limit its input in order to make rational sense, artificial intelligence in the form of neural network agents has no such limitations. More is better. If there is even the most remote possibility that the input may be relevant, it can be included in the data model used as input.

How does it work? How does the human body know to move the left toe or ankle first in order to deliver a punch with the right hand? That is certainly not a conscious effort on the part of the attacker. It is not necessary for us to know in order to move effectively and powerfully. Nor is it necessary for us to know in order to plan and execute an effective defense. The master of defense actually learns by acute observation and pattern recognition. The more knowledge we have of human anatomy and the laws of physics, the more quickly we learn. The beginner learns anatomy from the *Anatomy Coloring Book*, 159 pages of pictures and large print. The advanced martial artist has *Gray's Anatomy* on the book-shelf, 1257 pages with print small enough to overtax those with bifocals. And the master learns from using not only personal experience, but also the experience of others that came before. The master does not have to experience every possible combination of motions personally before antic-ipating their effect.

Just as Hanshi's knowledge and previous experience tells him certain early movement will result in some related subsequent move, predictive technologies also learn from experience. They must constantly monitor this massive collection of data in order to recognize patterns. They learn by recording events and analyzing patterns of data, which led up to their occurrence. By storing these massive collections of data, as a pattern of events or behavior begins to recur, the technology can predict what might happen next. The more closely the pattern matches that previously detected, the more accurate the prediction.

THE POSSIBILITIES

Patterned after more traditional uses of artificial intelligence, predicting line failures or when machines must be recalibrated is an obvious appli-cation of this technology. For many years, even those companies that implemented such predictive technologies operated them on the factory floor, in isolation from other information systems. Yet smart businesses don't need to stop here, but can continue to integrate these predictions with their business applications, which manage shop orders, to correlate potential unscheduled downtime with orders previously scheduled. In turn, the customers affected by the need to reschedule can be identified, possibly redirecting priorities in order that the negative impact to their business, and that of their customers, is minimized.

Other applications for neural network agents include the ability to detect or anticipate fraud or the recognition of patterns associated with check kiting or embezzlement. Or, the ability to flag potential credit risks once a certain pattern of ordering or payment (or nonpayment) occurs. Or it might be used to predict the fluctuations of currency or commodity pricing.

But the most obvious application of predictive technologies involves the forecasting of product demand. The advantage of a more accurate forecast is quite evident in the context of materials and production planning. However, the value of being able to predict buying behavior takes on a whole new dimension in the context of Internet-enabled selling. Anyone who has ordered a book from Amazon.com is familiar with Amazon's suggestions of other titles it thinks you might like. And everyone who has ordered a hamburger in a fast food restaurant has been asked, "Would you like fries with that?" Now, I'm not suggesting that Amazon.com is employing sophisticated predictive technologies and obviously the teenager taking your lunch order has simply been trained to suggest you leave a little more cash behind. Yet both of these are age-old selling techniques designed to get you to buy more product.

As you begin to computerize the buy–sell process, particularly with self-service order entry, there will be an increasing need to replace good old-fashioned salesmanship with more automated and electronic means of selling.

The more data that you can collect concerning buying patterns of your customers, the more effective you can be in predicting what promotions will be successful. Cross-selling and up-selling opportunities can be uncovered that go well beyond a retail environment. And as more and more companies seek to rationalize their suppliers to a trusted few, the more you can capture of your customers' spending, gain their trust and secure long-term relationships.

THE LIMITATIONS

However powerful these possibilities may seem, there is one sobering limitation to these technologies, and that is this: they are useless for predicting events that they have never observed. If the neural network agent has never "experienced" anything that produces the same pattern of events as a military conflict in desert conditions, a production line that has failed, or a consumer that bought fries with a burger, it has no "knowledge" of them and it cannot predict their effect on anything.

Today the application of predictive technologies to the general business world is in its infancy. As with any new application of technology, this calls for pioneering companies to roll out their own solutions. This means conditions and events must occur and be experienced at least once before the neural network agent can be of assistance in anticipating and analyzing them. Where these conditions and events represent seriously negative situations, this is obviously not an optimal solution, as the whole idea is to avoid them in the first place. The application of artificial intelligence will really come into its own when solution providers are able to offer "packages" of experience — pretrained neural network agents, which come with prepackaged experiences of other similar companies. This is exactly what technology companies have done in packaging artificial intelligence to monitor certain hardware and network configurations. So far, this same functionality remains a goal for even the most advanced software companies when applying this technology to business conditions and events.

But the potential does indeed exist in the future. Retail giants like Home Depot collect enormous volumes of information they are willing to share with suppliers and there is cross-product, cross-supplier and cross-industry potential in this data. Just as Hanshi is the "root master" of his arts, dedicated to the preservation of the teachings of those who have come before him, the most innovative software solution providers will become the "root masters" of businesses to come, preserving the lessons learned by pioneering users of predictive technologies.

SUMMARY

While speed is certainly a factor that influences success, in many ways, timing is far more important than speed. If your business is barreling down the business highway in the wrong direction, moving at warp speed will never get you where you need to go. As we learned in Chapter 2, leaders today cannot drive their businesses on historical data alone any more than the driver of an automobile can navigate only by monitoring the gauges on the dashboard and looking in the rearview mirror. Instead, drivers must also look ahead, collecting data from many sources, to anticipate conditions in the future. They must effectively use events of the past to predict events and conditions in the future.

8

EPILOGUE

The previous 12 months, the period of time during which I have written this book, have been some of the most turbulent times our economy has seen over the course of history. At the beginning of this period, in early July 2001, the stock market had rallied from its spring dip, although the "dot-com" frenzy had already turned into the "dot-com" deathwatch. While the ranks of the generation X millionaires dwindled, Wall Street began to value companies along the more traditional lines with which those of us in our forties and fifties were much more familiar and comfortable. Real assets and business models that strove for revenue and profits were finally coming back into focus as sound business plans. Having been thrown for a loop with market capitalizations that made you shake your head in wonder, it seemed like business life was finally getting back to "normal." It certainly seemed like an "optimistic businessperson" managing a traditional bricks and mortar business was no longer the oxymoron it had been for the past several years.

However, the rally of the stock market was very short-lived. Stocks began to plunge once more, taking the economy with them and, we thought, maybe optimism was premature. And then came September 11, 2001, a day, along with December 7, 1941, that will live in infamy forever. The very core of society in the United States was shaken and patriotism was revitalized. The travel industry took a hit that reverberated throughout many industries. The president declared war, but not against a sovereign nation. He declared war against terrorism, and many of our sons and daughters were called to duty. Many have faced combat. Some have died.

War typically rallies not only the troops, but manufacturing businesses as well, businesses that must increase production in order to supply the war effort. The ripple effect is usually sobering, yet economically positive.

And yet with this war, we still have not seen the turnaround in our economy that will signal optimism again.

This year was also marked with personal challenges for me. In April, interBiz, the division of Computer Associates for which I worked, was sold to SSA Global Technologies. Business applications had never been a focus or a strength for Computer Associates. The divestiture represented an effort to focus its energies on the core business. At the same time, the acquisition fit well into the SSA business strategy, which included growth of both market share and customer share. From a business standpoint this was a sensible and smart move for both companies. Yet obviously a transition like this is personally disruptive as corporate cultures and managements are merged. And it often means good people are necessarily made redundant. This was no exception and suddenly I found more than a few of my close friends out of work at one of the worst possible times to be job hunting.

I point this out only because it is representative of so many industries today. High tech is undergoing major readjustments as an industry right here and now. Corporate America has taken survival of the fittest to the next level and the leaner, meaner and flatter organizations are those that will thrive and survive. Yet that doesn't mean the development of new technologies has slowed. In fact, just the opposite seems to be the case. Gone are the distractions of the new business models that never touched products and never made a profit. Gone are the unrealistic expectations of the next "big thing," which generated a decade or more of double-digit growth for software companies, but resulted primarily in the stock-piling of shelf ware.

Information technology is getting down to business. I believe the next decade will bring the convergence of business and technology. Gone will be the elegant technical solutions in search of problems to solve. Pragmatic businesspeople will demand real solutions to real business problems. They will become responsible members of integrated business communities, and the price of admission to these will be shared information and interoperability.

ERP will play a key role in the continued success of enterprises worldwide. The integrated foundation of information ERP will provide, along with the transactional integrity it preserves, will be the underpinnings of E-business, which will be the next generation of business, as we know it today. That next generation will build upon that foundation to leverage all information assets available, to produce a heightened level of business intelligence. It will seek to integrate, not only software, but also companies, with shared information and shared services. Interoperability will become the fourth metric of success, along with price, quality and service.

So, at this time, as I draw to a close, economists are predicting recovery. While few business executives and managers are not quite optimistic yet, they are hopeful. The next generation of market leaders will emerge, and they will be the companies that have maximized their investment in ERP and have effectively transformed themselves for E-business profit.

Appendix

BUSINESS PROCESS OUTSOURCING

This appendix summarizes the results of a survey, conducted by AMR Research, Inc. (AMR). The survey was based on responses from 300 Chief Financial Officers (CFOs) from a variety of companies ranging in size and industry. Of these CFOs, all subscribers to *CFO* magazine, 62% represented companies with less than $1 billion in revenue, while the remainder were from larger companies. The results were reported in the January 2002 AMR Report, "Improving the Odds for Successful Business Process Outsourcing" by Lance Travis, Bob Kraus and John Hagerty. Excerpts and findings are included here with the permission of AMR.

AMR presents the bottom line conclusions from the study to be the following:

"Although most companies outsource complete business processes to a third party, it isn't a panacea for supporting all non-core business processes — poor service, unrealized savings, and hidden costs can kill the relationship.

"Business Process Outsourcing (BPO) — the contracting with a third party to provide complete staffing and support for a specific business process — is used by more than three fourths of companies with greater than $1B in revenue and more than two thirds of companies with less than $1B in revenue. More than 30 processes, ranging from travel handling and tax processing to manufacturing assembly and international trade services, are outsourced. However, consider the following:

- Outsourcing financial and administrative processes is more successful than outsourcing operational processes.
- Although the level of service received is not outstanding, 64% of users plan to expand their use of outsourcing."[1]

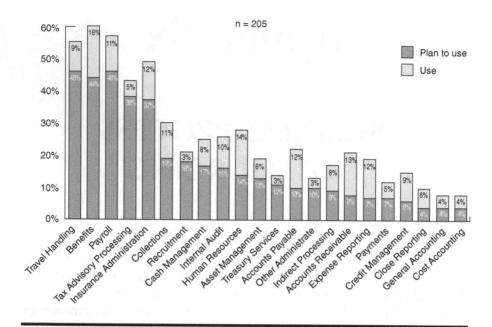

Figure A.1 Administrative processes outsourced. (Source: AMR Research, 2002.)

Figure A.1 depicts the administrative processes, while Figure A.2 displays the various operational processes being outsourced. Note that of all operational processes, third-party warehousing was the only one outsourced by more than 10%.

Figure A.3 indicates the level of satisfaction with various outsourced business functions. Despite CFOs being less than optimally satisfied, typically once the process is outsourced it is difficult to retreat from that position and return to bringing the work in-house, since in-house expertise is surrendered.

Figure A.4 indicates the most typical cause for dissatisfaction.

"A focus on core business processes drives many companies to consider BPO, but desired benefits are realized only when outsourcing finance and HR" (Human Resources).

"When asked about the factors that lead them to consider BPO, surveyed CFOs cite concentrating on core competencies, saving money, and tapping vendor expertise as the most important."[2] See Figure A.5.

According to the survey results, typically expected benefits had not been achieved with respect to call center operations, supply chain, logistics, and distribution operations.

"CFO should place more emphasis on domain expertise and SLAs (Service Level Agreements) during vendor selection to ensure quality service."[3] See Figure A.6.

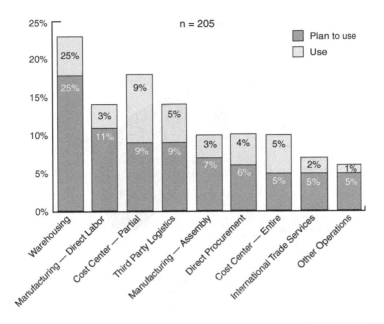

Figure A.2 Operational processes outsourced. (Source: AMR Research, 2002.)

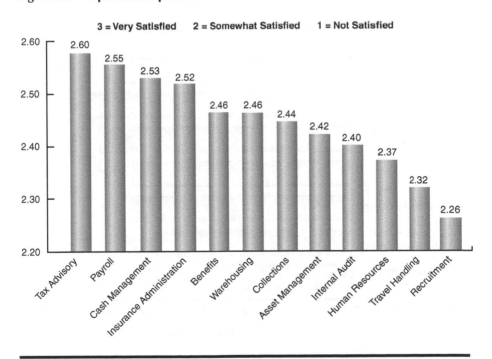

Figure A.3 Level of satisfaction. (Source: AMR Research, 2002.)

Percentage of Respondents

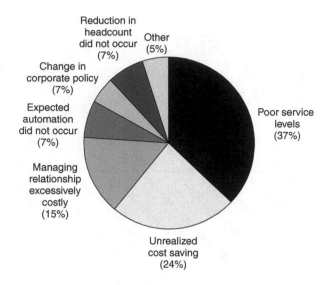

Figure A.4   Reasons for dissatisfaction. (Source: AMR Research, 2002.)

3 = Very Important   2 = Somewhat Important   1 = Not Important

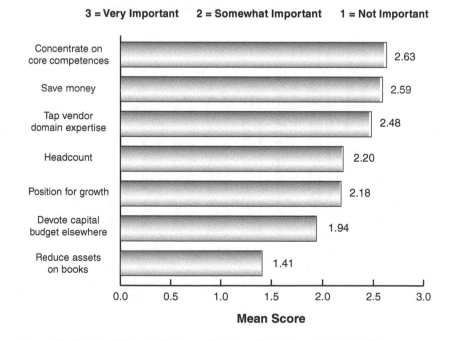

Figure A.5   Why consider BPO? (Source: AMR Research, 2002.)

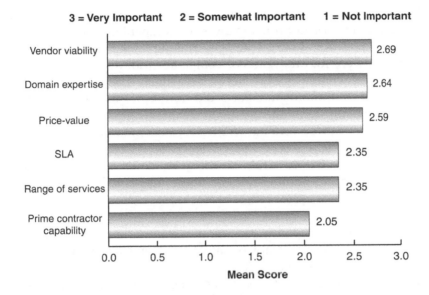

Figure A.6   Considerations in vendor selection. (Source: AMR Research, 2002.)

"CFOs prefer to pay on a transaction basis with multiyear contracts, but transaction costs may become prohibitively high and eliminate potential savings."[4] See Figure A.7.

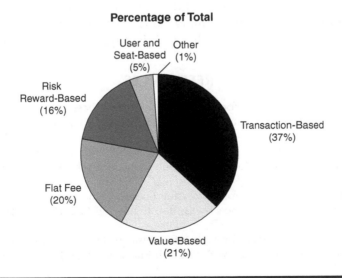

Figure A.7   Preferred payment methods. (Source: AMR Research, 2002.)

The survey also concluded that the length of service contracts varies most significantly based on company size. Companies with less than $1B in revenue opted for shorter contracts, limiting their exposure for failure. Fifty percent were for one year. For companies with revenues greater than $1B, three-year contracts were noted as optimal. Longer contracts were too restrictive, but shorter contracts did not allow sufficient opportunity for the relationship to mature.

As a result of these findings, AMR's recommendations to their clients are as follows:

"BPO has proven successful, especially in administrative areas such as payroll and insurance administration, but care needs to be taken to ensure success:

- Place more emphasis on vendor service levels during vendor selection. Vendor viability and cost savings are important, but performance is what will satisfy users.
- Include more than the CFO and the financial department during vendor assessments. Including line managers and the executives that own the processes being outsourced in the process will ensure that contractual SLAs will meet user needs.
- Create SLAs and performance metrics that are understandable and measurable; pay for performance and penalize for failure. Don't be one of the 43% of companies who do not perform periodic assessment of their BPO partner.
- Don't use outsourcing to abdicate the responsibility of cutting staff. Moving an underperforming department and its employees to an outsourcing firm will only result in paying more for bad service."[5]

REFERENCES

1. Improving the Odds for Successful Business Process Outsourcing, an AMR Research Report by Lance Travis, Bob Kraus, and John Hagerty, January 2002, p. 1.
2. Improving the Odds for Successful Business Process Outsourcing, an AMR Research Report by Lance Travis, Bob Kraus, and John Hagerty, January 2002, p. 6.
3. Improving the Odds for Successful Business Process Outsourcing, an AMR Research Report by Lance Travis, Bob Kraus, and John Hagerty, January 2002, p. 8.
4. Improving the Odds for Successful Business Process Outsourcing, an AMR Research Report by Lance Travis, Bob Kraus, and John Hagerty, January 2002, p. 9.
5. Improving the Odds for Successful Business Process Outsourcing, an AMR Research Report by Lance Travis, Bob Kraus, and John Hagerty, January 2002, p. 13.

INDEX